Light: Neither a Wave, Nor a Photon

Sandeep Chavan

Published by SANDEEP CHAVAN, 2025.

LIGHT: NEITHER A WAVE, NOR A PHOTON

First edition. August 9, 2025.

ISBN: 979-8231090211

Written by Sandeep Chavan.

Table of Contents

To **Light**—Not the wave they imagined, nor the photon they defined,but the silent messenger that has always been here,revealing the universe,and yet keeping its own truth hidden.

"You never saw light. You saw the universe looking back at you."

— Sandeep Chavan

Disclaimer

This book presents an alternative perspective on the nature of light, developed under the framework of Universal Energy Dynamics (UED). The concepts, explanations, and terminology herein are the result of independent research and original interpretation by the author. While every effort has been made to ensure accuracy and coherence, some ideas challenge or diverge from established scientific theories and may not yet be experimentally verified or accepted by the mainstream scientific community.

The content is intended for educational and exploratory purposes only. Readers are encouraged to critically evaluate the material, cross-reference with other sources, and draw their own conclusions. The author and publisher disclaim any liability for the use, misuse, or interpretation of the information provided in this book.

Opening Note

The Illusion That Lit the World

"Before you saw anything, you were already present. And when you looked, the universe aligned itself and called it light."

We have spent centuries chasing light.

We've dissected it into waves and photons, wrapped it in theories of duality, trapped it in equations, bent it around stars, and fired it through slits and mirrors. We've built machines to measure its speed, tools to harness its force, and philosophies to marvel at its mystery.

And yet... we never truly asked:

What is it that we're actually seeing?

Is light a thing? A traveler? A messenger? A bouncing beam?

Or is it simply a consequence—a moment of ripple alignment—arising the instant your intent breaks the silence of the field?

In this book, we won't treat light as a substance or a wave. We won't pretend it moves. We won't bow to photons or speed limits. We'll dare to look beyond the illusion.

We'll show that light is not what travels, but what happens—at the very moment you choose to observe.

We'll see that the universe doesn't shine on its own. It responds. It resolves. It aligns.

And that alignment—that first event—is what you call light.

This book is not about understanding optics. It's about undoing the false clarity that has blinded our vision.

It's time to return to the source—not of illumination, but of meaning. Where nothing moves, yet everything appears. Where silence gives rise to sensation. Where light is not out there, but within the alignment you create.

Welcome to the still universe.

Let us begin.

Preface

From the first time you looked up at the night sky, you were taught a story: that light travels across unimaginable distances, carrying messages from the past; that stars move, galaxies expand, and the universe is in constant motion. You were told these as facts, and you believed them — as did I, for many years.

But belief is not the same as knowing. And knowing is not the same as seeing what is truly there.

This book was born out of a quiet rebellion against unquestioned truths. Not a rejection for the sake of contrarianism, but a patient dismantling of the assumptions we inherited about light, space, and motion. In their place emerges something startlingly simple — a universe that has never moved, a reality that exists as a still field, and "light" not as a traveler but as the handshake of alignment between observer and observed.

You will find here no allegiance to mainstream cosmology, nor a call to abandon it in blind faith of something else. Instead, you will find a set of tools, thought experiments, and perspectives drawn from Universal Energy Dynamics (UED) — a framework that asks us to look again at the fundamentals, without the comfort of the stories we have told ourselves for centuries.

This journey is not about light alone. It is about you, the observer. For if the universe is still, then your perception, your alignment, and your intent become the true architects of what you call reality.

As you turn these pages, I ask only this: pause often. Question relentlessly. Let go of the need for the universe to behave like a moving machine. And be open to the possibility that stillness — not motion — is the truest form of cosmic elegance.

The universe has been waiting for you to notice.

— *Sandeep Chavan*

How to Read This Book

This is not a book you simply read from start to finish — it's a field you step into. The ideas here are not meant to be absorbed passively; they are meant to be tested, questioned, and felt.

You will notice three distinct rhythms in the chapters:

1. **Revealing the Illusion** – These sections dismantle long-held assumptions about light, motion, and the universe. They may contradict what you've been taught. Read them slowly. Let the discomfort in, but resist the urge to defend old ideas too quickly.
2. **The UED Perspective** – Here, Universal Energy Dynamics reframes what you just deconstructed. You'll see terms like SVE (Structured Vacuum Energy), ATT (Alignment Threshold Time), ripple delay, and alignment zones. Don't rush past them. These are not "alternative jargon" — they are the coordinates of a different way of seeing.
3. **Implications and Tools** – These parts bring the discussion back to you: the observer. How your perception shapes reality. How stillness changes your role in the cosmos. How technology, philosophy, and self-awareness all shift when you change the frame of reference.

Reading Tips

- **Pause at the metaphors** – They are not decoration. They're meant to make abstract concepts real enough to hold.

- **Don't get stuck on the math** – When equations appear, see them as signposts, not barriers. They show that this framework is testable, but you don't need to solve them to grasp the point.
- **Move between chapters freely** – While the book flows in sequence, each chapter stands on its own. If a topic intrigues you (say, "Cosmic Stillness"), you can jump there directly and still follow the argument.
- **Use the Appendices** – The comparison tables, glossary, and experimental suggestions at the back are there to anchor you if a concept feels unfamiliar.

Finally, understand that this book is not asking you to believe in a new model of reality. It is asking you to stand in stillness long enough to see what has been there all along. If you do, you may never see light — or yourself — in the same way again.

Prologue

The Day the Universe Looked Back

He opened his eyes, but he didn't see the light.

He saw the ripple.

It was not a beam. Not a ray. Not even a glow. It was a feeling—a subtle shift in the silence that surrounded him. A crack in stillness. A pulse that didn't come from outside, but from somewhere deeper. It was as if the universe... paused. And in that pause, it aligned itself with his attention.

He had no word for it. Not yet. So he called it light.

Not because it lit the room. But because it lit the meaning of being aware.

We think light is something out there. We believe it travels. Bounces. Reflects. Warps around stars and pierces the darkness. But what if none of that is true?

What if light is not a traveler, but a trigger? What if it's not something emitted or received, but something arising—as a consequence of your presence?

You don't see because light enters your eyes. You see because you exist as an observer field capable of ripple alignment.

That's the secret we missed for centuries.

We kept measuring brightness. We never questioned why the universe becomes visible only when we choose to look.

This is not a book about optics. This is a book about the first event that occurs every time silence is disturbed by intention.

It is about a universe that doesn't glow on its own, but responds to attention. About a light that is not sent, but converged. About seeing not as a passive act, but as a field-level resonance—where reality becomes measurable only because you were there to give it meaning.

So let us begin from where it all truly starts: Not from the Sun. Not from a photon. Not from speed.

But from the stillness that broke the moment you decided to look.

And in that instant—light was born.

Part I: Breaking the Illusion

Chapter 1: The Greatest Misunderstanding

How humanity got light wrong—waves, photons, and everything in between

"It was never light that confused us. It was our refusal to accept that seeing might not require something to travel."

We built entire civilizations around the glow of the sun, told stories of gods descending in beams of light, and later, drew elegant wave equations and invented mysterious particles to explain what was really just a response to attention.

We called it a wave. Then, we called it a photon. Then, when both failed, we called it both. And when neither made full sense, we told ourselves: "Well, that's quantum."

But the truth is simpler. More elegant. More unsettling.

We never actually saw light. We saw a field adjusting itself to our presence. We saw the universe resolving itself into visibility—because we looked.

And instead of recognizing that, we gave it a name, a speed, and a set of rules—rules that made sense only if we agreed to ignore the stillness behind it all.

This chapter is not here to update your physics textbook. It's here to burn the textbook.

Let's begin with the illusion we've called light.

1.1 – The Beginning of Bright Ideas

How the reverence for light shaped our deepest intuitions—and our earliest mistakes.

Long before science tried to explain light, humans bowed to it. We worshipped it. Feared it. Assigned it the role of gods and messengers. Because nothing stirred the soul—or the imagination—like that which broke the darkness.

Fire was the first miracle. The controlled glow of a wild, untouchable force. It brought warmth, vision, and safety in a chaotic world. No wonder it became sacred. The Sun, rising and setting with unwavering rhythm, became the face of divinity—an eye in the sky that could give life or take it.

In nearly every ancient tradition, light was synonymous with truth, clarity, awakening. From the Vedic "jyoti" to the Greek "phōs," from Egyptian sun gods to Persian fire rituals, illumination was not a physical process—it was a spiritual one.

But here lies the paradox:

We revered light as if it were divine—but we tried to explain it as if it were material.

Light as Messenger: The Philosophical Divide

The earliest thinkers didn't agree on what light *was*, but most agreed it must be something *in motion*.

Greek philosophers debated whether vision occurred because light came out of the eye (the *emission theory*) or because light entered the eye (the *intromission theory*).

- Empedocles believed the eye emitted rays that touched objects.
- Later, thinkers like Aristotle argued that objects emitted something received by the eye.

Ancient Indian thinkers, especially in the Nyaya and Vaisheshika schools, also explored perception:

- They postulated light as a substance, composed of minute particles (Tejas), which traveled from the object to the observer.
- The concept of *Prakash* (illumination) was closely tied to cognition, but always required transmission.

Whether East or West, the assumption was shared: For vision to occur, something must leave the object, or the eye, or both—and move.

The Trap of Intuition

To be fair, this assumption felt natural. If you see a torchlight beam in the fog, or sunlight piercing clouds, it *looks* like something is moving. If you close your eyes, vision stops—surely something *must* be arriving?

And that's the trap.

Humanity assumed that what is experienced must be transmitted. That what is seen must have traveled to be seen.

No one asked the deeper question: What if vision doesn't come from reception, but from alignment? What if light is not a traveler, but a *response*?

Why This Matters

This ancient framing of light as a carrier—a traveler that brings truth from one place to another—became the bedrock for all future science.

And yet, it was based on nothing but:

- Observation of surface-level phenomena
- Intuition rooted in survival (dark = danger, light = safety)
- Symbolic associations (light = good, dark = evil)
- And the simple, unquestioned belief that something must

move for something to be known

But in doing so, we locked ourselves into a false framework.

We started explaining a still universe using ideas of motion. We turned the first event of ripple alignment into a story of travel and speed. And we've been correcting and patching that misunderstanding ever since.

Section Summary:

- Light was always sacred—but never understood.
- From myth to philosophy, it was treated as a transmitter, not a consequence.
- The idea that something must move to be seen became the first great error in the understanding of light.
- And this error became so deeply embedded, we built all modern optics around it—without ever revisiting the question of what actually happens when we see.

1.2 – The Wave Age: A Smooth but Shaky Story

How a graceful theory gave us partial answers—and permanent confusion.

After centuries of debating whether light traveled out of the eye or into it, science finally made a breakthrough—or so it seemed.

In the 17th century, Christiaan Huygens proposed that light was a wave, not a particle. He described it as spreading outward like ripples in a pond, traveling through a medium just like sound through air or water.

It was a clean, elegant model. Wave behavior explained a lot: reflection, refraction, and even interference. It felt like the dawn of clarity. Light wasn't mystical—it was mechanical.

But one question lingered like a silent ghost:

A wave of what?

The Aether Fantasy

For any wave to propagate, it must have a medium.

- Sound waves need air.
- Water waves need water.
- Seismic waves need the ground.

So scientists imagined a universal substance that filled all space—one that allowed light to ripple through even the vacuum between planets and stars.

They called it the aether.

This mysterious medium was said to be:

- Weightless, invisible, frictionless
- Perfectly elastic, yet undetectable
- Present everywhere, yet affecting nothing else

It became the greatest assumption science ever clung to without evidence.

The aether was the silent partner of wave theory—never seen, never felt, never measured. But it *had* to be there... because how else could light move?

Maxwell's Equations and the Rise of EM Waves

Then came James Clerk Maxwell in the 19th century, who formulated a stunning set of equations uniting electricity and magnetism. His math predicted that electromagnetic waves—oscillating electric and magnetic fields—should propagate at the exact speed of light.

Boom. Light was now an electromagnetic wave. It no longer needed to be a "thing"—just a self-propagating vibration through space.

The wave theory gained prestige, legitimacy, and mathematical muscle. It became the mainstream model. Light was declared victorious as a transverse wave.

But again... a quiet problem remained.

The Aether's Silent Death

In 1887, Michelson and Morley conducted a now-famous experiment to detect the motion of Earth through the aether. They expected to find differences in light's speed depending on direction of travel—just like wind affects a boat's speed.

They found nothing. No wind. No drag. No preferred direction. No aether.

The very medium that justified the wave model of light... simply wasn't there. And yet, instead of rethinking the wave model, science discarded the medium and kept the wave.

This was the beginning of an uncomfortable contradiction:

- A wave... with no medium
- A propagation... with no substance
- A vibration... of what, exactly?

No one could say.

The Patchwork of Phenomena

Still, wave theory *worked*—or at least, it explained enough:

- Interference patterns in the double-slit experiment
- Diffraction through slits and edges
- Polarization of transverse waves

These were real, measurable effects—so science doubled down on the model.

But it became increasingly clear that wave theory couldn't explain everything:

- Why did light sometimes behave like a particle?
- How could it propagate through absolute vacuum?
- What caused a wave to "collapse" when observed?
- Why did light not slow down or decay like other waves?

Every answer led to another patch. Every patch led to more contradictions. And still... no one dared to say what was becoming obvious:

Maybe light wasn't a wave at all. Maybe it was just a consequence of something deeper—something that never moved.

Section Summary:

- Huygens and Maxwell offered elegant frameworks—but only within a flawed premise: that light must be moving.
- The wave model explained many effects, but relied on the unproven existence of a medium (aether).
- When the aether died, wave theory was kept alive on faith and math, not on foundational clarity.
- Even its most impressive explanations were interpretations, not answers.
- The greatest irony? We created a mechanical model to explain a non-mechanical phenomenon.

Light behaves like a wave only when we expect it to. But it has never once told us what it truly is.

1.3 – Photon Fiction and the Duality Patch

The quantum 'solution' that only deepened the mystery.

By the dawn of the 20th century, the wave model of light—though elegant—was quietly falling apart. The aether was dead. Light behaved like a wave in some experiments, but not in others. And worst of all, some phenomena just wouldn't play by the wave rulebook.

One of those phenomena broke the illusion wide open: The Photoelectric Effect.

Einstein's "Miracle": A New Patch, Not a New Truth

In 1905, Albert Einstein published a radical interpretation of an unresolved observation:

When light strikes certain metals, it kicks electrons out of their surface. But strangely, increasing the light's intensity didn't increase the energy of the ejected electrons. Instead, only the frequency of light mattered. Even a dim beam of high-frequency light could knock out electrons, while a bright low-frequency beam could not.

The wave model couldn't explain this.

Einstein proposed something wild (at the time):

Light comes in discrete packets of energy. He called them quanta. Later, we called them photons.

Each packet carried energy proportional to its frequency:

$E=hf$

Where E is energy, h is Planck's constant, and f is frequency.

This solved the photoelectric puzzle. It was hailed as brilliant. But underneath that equation, we quietly shifted the narrative: Light was now a particle. A traveling energy bullet.

The Rise of the Photon – and the Fall of Understanding

The photon was a convenient invention—too convenient. It let scientists imagine a particle of light, something that:

- Carried energy across space
- Could collide with matter
- Could behave like a classical object
- Could be *measured* and *counted*

But this created a new contradiction:

- In some experiments, light behaved like a wave
- In others, like a particle

And instead of questioning the foundation, physics created a new doctrine: Wave-particle duality.

Wave-Particle Duality: The Diplomatic Lie

Rather than choosing between two incompatible models, we declared both correct.

"Light is a wave *and* a particle... depending on how you look at it."

This was not a breakthrough. It was a surrender.

We stopped asking what light *is*—and settled for how it *behaves under measurement*. We accepted contradiction as reality. We wrapped mystery in math—and moved on.

The Double-Slit Deception

The famous double-slit experiment was the ultimate showcase of duality. Fire photons through two slits one at a time, and over time, they interfere with themselves, forming a wave-like pattern.

But when observed directly—boom!—the interference vanishes. The photon behaves like a particle again.

Physicists were stunned. They blamed observation for collapsing the wave. But they never fully asked:

What is observing? Who is the observer? What changes in the field when intent enters the equation? What if the particle never existed—and all we saw was ripple alignment responding to attention?

Instead, they rebranded the paradox: "It's quantum."

And the photon survived.

The Real Question We Refused to Ask

We named it a photon. We assigned it a speed. We tracked its energy. We built detectors to "count" its arrival.

But we never answered:

- What is actually *moving*, if anything?
- What is a photon *made of*?
- Why does it know whether it's being observed?
- How does it interfere with itself if it's a particle?
- What exactly is being observed—and through what mechanism?

These questions remain open because we never questioned our assumption that something had to *travel* in the first place.

We invented the photon to preserve that illusion.

Section Summary:

- Einstein's photon was a clever fix for a specific problem—not a complete model of light.
- Wave-particle duality was not an answer, but a compromise that avoided deeper inquiry.
- The observer's role in experiments like the double-slit was never truly explored—only *mathematically absorbed.*
- The photon kept science moving forward technologically, but philosophically, it marked a dead end.

"The photon gave us numbers, but stole the truth. We didn't understand light—we just got better at ignoring the mystery."

1.4 – The Constant Speed Lie

Why "c" has nothing to do with speed—and everything to do with stillness responding.

If the photon was the bandage, then "c" became the plaster that held the illusion together.

After Einstein's 1905 proposal that light behaved like discrete energy packets, he made another, bolder leap:

The speed of light in vacuum is constant, for all observers, in all frames of reference.

This was revolutionary—and foundational. From it emerged Special Relativity, time dilation, mass-energy equivalence, and the very idea that nothing could ever exceed c.

But what if we told you... Light never had a speed.

And what Einstein called "constant" was not motion—but a misunderstood ripple response of a still universe?

Light Appears Fast. But Does It Actually Move?

We assume light moves because:

- It arrives after a delay
- It creates shadows and beams
- We can measure time between emission and detection

But this delay is not travel. It is ripple re-alignment across the SVE (Structured Vacuum Energy)—a *response*, not a transmission.

What you measure as light "arriving" is the field realigning itself to your observer-node based on ripple tension.

So the "speed" of light is really:

The maximum rate at which ripple tension can resolve across the field.

This rate appears constant not because light is consistent in motion, but because the universe itself is consistent in how it responds.

Einstein's Leap: Brilliant, but Incomplete

Einstein's declaration that c is invariant was a powerful move—and mathematically successful.

It explained:

- Why different observers measure the same light speed
- Why time and space must flex to maintain that constancy
- Why light sets the cosmic speed limit

But there was a fatal assumption buried in it:

That light moves.

Einstein never questioned *whether* light was moving—only how fast it seemed to.

He built a beautiful geometry around that assumption. But he never asked:

- What is actually moving?
- Is there even a "thing" in motion?
- What if the delay is not displacement—but field recalibration?

Relativity is a masterpiece—but it is not the truth. It is a brilliantly crafted response to a false premise.

Measuring Ripple Delay as Velocity: A Category Error

All our light-speed measurements are based on:

- Emission at point A
- Detection at point B
- Time difference = speed

But under UED (Universal Energy Dynamics), this logic collapses:

- No photon traveled
- No wave propagated
- What occurred was ripple tension resolution across SVE, reaching the observer's ATT (Alignment Threshold Time)

What you measure is not transit time—but alignment delay. And the distance is not spatial—it is a ripple gradient.

So:

Speed of light = (Ripple Resolution Distance) / (Ripple Response Time) Which is not speed at all, but a constant field reaction rate.

In UED terms:

$$c = \left(\frac{\Delta R_{align}}{\Delta \tau_{field}} \right)_{max}$$

Where:

- ΔR_{align} = Ripple alignment range
- $\Delta \tau_{field}$ = Minimum ripple response time of SVE

The Blinding Assumption: That Light Must Move

Why are we so attached to the idea that light travels?

Because we see beams. Because we measure delays. Because it *feels right.*

But this attachment blinds us to a deeper truth:

The universe is not made of things moving. It is made of ripple fields resolving into form whenever and wherever an observer intends.

The UED Pivot: Light is Not an Entity. It's a Consequence.

This is the moment the illusion collapses.

Everything you thought about light—its speed, motion, energy, behavior—is the after-effect of ripple alignment.

- No particle traveled.
- No wave moved.
- The universe didn't emit anything—it responded to your alignment.
- What you call "light" is the first event in the chain of perception.

And the "speed of light" is simply the limit of how fast the field can respond to ripple misalignment.

No more. No less.

Section Summary:

- The constancy of light's speed is real—but it's not a speed.
- Einstein explained the consequences, but never questioned the core premise of motion.
- All measurements of light delay are misread ripple response timings.
- UED redefines c not as velocity, but as maximum ripple resolution rate across SVE.
- The idea that "light moves" was the last lie holding the illusion together.

The speed of light is not the speed of a thing. It is the rhythm of a universe remembering how to be seen.

A Reframing of Reality

We didn't discover light. We didn't decode it. We didn't even observe it.

We named it—and that gave us the illusion that we understood it.

We invented waves to describe its elegance. We invented photons to explain its power. And when neither could hold the full truth, we invented a doctrine of contradiction and called it *duality*.

We wrapped light in equations, defined its speed, bent time and space around it. And yet, we never once stopped to ask the most important question:

What exactly are we seeing—and why does it appear in the first place?

The answer has been hiding in plain sight, not in the particle, not in the wave, but in the ripple consequence that occurs the moment intent meets silence.

Light isn't something we chase. It isn't something we catch. It isn't something that travels across the void to reach us.

It is the void responding. It is stillness breaking. It is the first ripple of a universe aligning itself with your presence.

So no—light was never misunderstood as a wave or a particle. Light was never a "thing" at all. It was a silence misinterpreted.

And we, desperate to make sense of the glow, built an entire worldview on that misinterpretation. But now, we let the illusion dissolve. Not to discard light—but to finally see what it actually is.

Not an object. Not a traveler. But the first event of alignment. And in the chapters ahead, we will go where science hesitated:

Into the stillness behind the shine. Into the field beneath the flash. Into the truth that light was never seen—it was always resolved.

Let there be ripple.

Chapter 2: The Myth of Speed and Travel

Why light doesn't travel—and what "c" actually means in a still universe

"The moment you let go of movement, the entire universe becomes understandable."

Everything we know—science, measurement, progress, even survival—has been built on the assumption that things move.

Arrows fly. Planets orbit. Winds rush. Blood flows. And light—our most sacred messenger—is believed to travel faster than anything else in existence.

So when we see delay, we say something must be in motion. We measure the time between flash and detection, and we declare a speed. We call it c—the speed of light. We build theories around it. We bend time and space to preserve it. We crown it the king of constants.

But here's the paradox:

What if nothing is actually moving at all? What if light doesn't travel—because there's no need for anything to travel in a universe that is already fully connected and fundamentally still?

What you call "speed" may just be how fast the universe can respond to your intent. And what you call "light" may just be the alignment of stillness around your awareness.

This chapter will undo the mythology of speed. Not to slow things down—but to show you how motion itself is an illusion, born from ripple tension resolving across a silent field.

We are not chasing light. We are watching the field resolve itself around us—and mistaking that elegant recalibration for velocity.

Let us step beyond the idea of transit.

Let us meet the still universe.

2.1 – The Speed Trap

Why we see motion even in stillness—and how it blinded us to the true nature of light.

Since the earliest days of human observation, motion has been our obsession. It is the clearest marker of change, threat, or opportunity. A rustle in the bushes. A galloping horse. A falling apple. A shooting star.

To move is to act. To act is to exist. To not move, for the human mind, has always been equivalent to death, emptiness, or irrelevance.

So we built our understanding of the world around the concept of movement.

- Objects move through space.
- Forces cause motion.
- Time progresses by the movement of hands on a clock or stars in the sky.

And when it came to light, we followed the same pattern.

From Arrows to Beams: The Continuity of Assumption

When an archer releases a bowstring and the arrow strikes a target, the mind sees a simple chain:

Cause → Motion → Effect.

That same logic was applied upward through the centuries:

- Light from the sun? Must be flying particles or traveling waves.
- Fire illuminating a wall? Must be emitted rays.
- Flashing lights from distant stars? Must be crossing the void.

We assumed delay = distance ÷ speed. And any delay in perception must imply something traveled to reach us.

We didn't question the premise—we just measured it better and better. Eventually, we measured this so-called "speed of light" to absurd precision:

$$c=299{,}792{,}458 \, m/s$$

But we never asked:

What exactly is moving at that speed?

Not a particle. Not a wave. Not a substance.

Just... a consequence.

Delay ≠ Travel

The human mind evolved to interpret time gaps as motion. This makes sense when watching falling objects or hearing thunder after lightning.

But it breaks down when applied to light.

Here's why:

- Thunder has a known source (lightning) and travels through air.
- Light is assumed to travel similarly—across space from source to observer.
- But space, unlike air, is not a medium with known tension or structure in classical physics.
- We detect delay, and we map it as speed—but what if that delay is simply ripple resolution time, not transit time?

Just because perception is not instantaneous doesn't mean something crossed space.

It may simply mean that the field is still—and your observation took time to trigger the first ripple of alignment.

Speed as a Comfort Mechanism

Our belief in speed gives us order. If something moves, we can track it, predict it, calculate its impact.

That's why "speed" became the default explanation for every delay-based phenomenon:

- Light arriving from the sun? Speed.
- Radio signals from space? Speed.
- Laser beam detection? Speed.

We layered precision on a flawed foundation—never questioning the premise.

Speed was easier to accept than ripple reconfiguration in a still field.

Because speed gives you the illusion that:

- Something real moved
- Something left and arrived
- Something happened "out there"

But what if nothing left? What if the event you witnessed was always here, and it was your intent that aligned the field enough to make it visible?

Light: The Final Illusion of Motion

Light became the last untouchable domain of speed-based thinking.

- You can slow it down in materials.
- You can bend it with lenses.

- You can bounce it off mirrors.
- But in vacuum, it's always measured as moving at c.

And so, we concluded:

Light must be moving—even when we don't know what it is.

But in Universal Energy Dynamics (UED), we propose a radical alternative:

Light does not move. The field realigns. And what you measure as speed is simply how fast ripple tension reaches your ATT (Action Trigger Threshold).

Section Summary:

- Human perception instinctively equates delay with motion—and science followed that path.
- "Speed" became our psychological and mathematical crutch for understanding cause and effect.
- But light's constancy and delay do not prove travel—only that a structured response is happening.
- Light is the ultimate illusion of movement—a ripple consequence misread as a journey.

"We chased light across space, only to find that we never moved, and neither did it. The only thing that changed was alignment."

2.2 – What Really Happens: Ripple Response, Not Transit

Light never travels. The universe simply realigns around your intent.

If motion is an illusion, then what *actually* happens when we "see" light? When we strike a match, or gaze at a star, or flick on a flashlight—what's really going on?

In the classical view, the explanation is simple: Something (a photon or wave) leaves the source, travels through space, and eventually arrives at your eye.

But in the Universal Energy Dynamics (UED) framework, this is completely false.

There is no traveler. There is no crossing of space. There is no object making a journey.

Instead, what occurs is far more subtle—and far more profound.

You don't receive light. You align with the ripple that makes it visible.

Let's unpack this.

The Universe Is Not Empty — It Is Structured

What we call "vacuum" is not empty. It is Structured Vacuum Energy (SVE)—a still, fully connected, undivided field that underlies all phenomena.

SVE is not space. It is not void. It is not made of particles or waves. It is the background equilibrium—the infinite silence from which all ripple emerges.

Think of it not as a stage on which things happen, but as the substance of reality itself, before any form takes shape.

Everything you observe—light, sound, matter, force—is simply a ripple or a tension pattern that temporarily disturbs this perfect stillness.

So What Is a Light Event?

A light event is not something sent or received.

It is a reconfiguration of ripple tension in the SVE, triggered by interaction between two field zones: the source node and the observer node.

When a field disturbance (say, an atomic transition) occurs at the source:

- It doesn't emit a particle.

- It doesn't push out a wavefront.

Instead:

It creates a ripple tension differential in the SVE.

This ripple propagates—not through motion, but through alignment pressure, like concentric ripples nudging the field.

And when that ripple pattern resolves at your observer node (your eye, detector, or sensor), it crosses a specific threshold.

That threshold is called the Action Threshold Time (ATT).

The Role of ATT: Where Light "Begins" for You

In UED, nothing becomes observable until the ripple tension:

- Reaches the observer node, and
- Crosses the ATT—the minimum ripple tension required to trigger perception or detection

Only then do you say, "I saw light," or "I received a photon."

But the truth is:

- Nothing arrived.
- Nothing was sent.
- The field simply reconfigured itself around your alignment.

The delay you measure is not transit time. It is ripple resolution time.

The Local Realignment of a Still Universe

To simplify:

- A candle flickers.
- Ripple tension emerges from that disturbance in the SVE.
- Your field, as an observer, is already connected to the same SVE.
- When the ripple realigns at your ATT, you experience

"light."

There was no transfer. No crossing of space. Only coherence forming in a shared medium.

This is why, no matter how far the source is, light appears to "move" at the same constant speed.

You are not measuring motion. You are measuring the time it takes for ripple to reach perceptual resolution.

Why It Looks Like Travel

Our minds interpret this delay as:

- A photon in flight
- A beam in motion
- A wave crossing space

But in truth:

- Space was never crossed.
- The field was never broken.
- The observer and the source were already linked through SVE.

It is like pressing a key on a piano in a quiet room. The sound doesn't travel to your ears. The entire room responds—tension realigns—and you hear the note.

In the case of light:

The field realigns. The ripple resolves. The silence becomes visible.

That's what we mistake for light "arriving."

Section Summary:

- Light does not travel through space—it is a ripple resolution across SVE.

- There is no particle or wave making a journey—only alignment between field nodes.
- The Action Threshold Time (ATT) is what determines when a ripple becomes observable.
- The delay we measure is not movement—it is field response time, creating the illusion of speed.

"You are not seeing what moved. You are seeing how the field resolved itself into form—because you were there to see it."

2.3 – The Misreading of "c"

Why the so-called speed of light has nothing to do with speed—and everything to do with ripple response.

For over a century, the value of c—299,792,458 meters per second—has held a sacred position in physics. It defines the upper limit of velocity in the known universe. It anchors Einstein's theory of relativity. It appears in equations across quantum mechanics, electromagnetism, and cosmology.

But beneath its precision lies a critical flaw:

We assumed that c represents a speed because we assumed that light moves.

In the UED framework, this assumption is dismantled entirely.

Light Doesn't Move. So What Is c Measuring?

As established earlier, in Universal Energy Dynamics (UED):

- Light is not an entity.
- It does not travel.
- What we experience as "light" is a ripple reconfiguration across the Structured Vacuum Energy (SVE) field.
- It becomes perceivable only when that ripple crosses the Alignment Threshold Time (ATT) at the observer node.

So what, then, is c?

c is not a velocity. It is the maximum rate at which the ripple alignment can resolve across SVE. It's not about travel—it's about response.

Why c Appears Constant for All Observers

In Einstein's special relativity, c's constancy—regardless of the observer's frame of reference—was astonishing. But in UED, it's not astonishing at all.

It's inevitable.

Why?

Because:

- All observers exist within the same SVE framework, which is inherently still and continuous.
- The ripple propagation is not constrained by relative motion, but by the field's intrinsic coherence.
- Ripple resolution across the field happens at the same rate, independent of spatial reference frames.

It's like tuning multiple instruments to the same silent string. Regardless of who plucks it—or where—it vibrates at the same rate because it's connected to the same fundamental stillness.

Thus:

What Einstein called "light speed" is really the field's ripple equilibrium limit—a maximum rate of alignment within SVE, not a particle's cruise control.

Ripple Resolution, Not Particle Velocity

Let's redefine c in UED terms:

$$c = \left(\frac{\Delta R_{align}}{\Delta \tau_{ATT}} \right)_{max}$$

Where:

- ΔR_{align}: Ripple alignment range in the field
- $\Delta \tau_{ATT}$: The minimum Alignment Threshold Time required for resolution
- The result is not speed through space, but the rate at which ripple alignment can occur across SVE

There is no entity moving from point A to point B. What changes is the state of the field—from misaligned to aligned—perceived by the observer when ATT is crossed.

This rate of resolution appears *like* a speed because we misinterpret ripple delay as travel.

The Uniformity of SVE

Another reason c appears universal is because SVE itself is uniform—not physically, but functionally.

Even though local ripple configurations can vary (causing lensing, slowing of light in glass, etc.), the core resolution rate of ripple alignment in SVE remains bounded.

This is why:

- Light appears to slow in dense media (because ripple coherence is disrupted), but never exceeds c.
- c is the upper limit, not because of motion, but because it is the maximum allowable ripple resolution rate before decoherence.

No ripple can realign faster than the field can process equilibrium. That's what c really is.

Why the Velocity Interpretation Fails

By interpreting c as a velocity:

- We assume transit, where there is only transition.
- We create illusions of photons in flight, spacetime warping, and particles hitting targets.
- We overcomplicate reality with relativistic mechanics—when the core truth is that the field is still, and c is simply its ripple response limit.

UED replaces this with a stillness-based model:

- No movement
- No transmission
- Only ripple tension resolving across an infinite, coherent medium

Section Summary:

- The traditional interpretation of c as the "speed of light" is a category error—confusing ripple response time with velocity.
- In UED, c is redefined as the maximum rate of ripple alignment across the SVE.
- Its constancy is not mysterious—it's the natural behavior of a fully connected, ripple-responsive field.
- Light doesn't travel at c—the field resolves your alignment at c.

"c is not how fast light moves. It is how quickly stillness becomes visible when you are ready to see."

2.4 – Implications of Stillness

What happens when you stop believing in movement—and start seeing ripple.

If the universe is fundamentally still—truly still—then everything we thought we knew about motion, space, time, and causality begins to unravel.

This isn't just a theoretical idea. It's a complete shift in perception.

Because if nothing actually moves, then:

- Nothing travels.
- No force pushes.
- No signal is transmitted.

Instead, all that exists is alignment and misalignment within a perfectly still, ripple-responsive field.

This is the implication of stillness in Universal Energy Dynamics (UED).

Let's walk through what this means—technically, observationally, and philosophically.

Distance Is Not Separation — It's Ripple Differential

In classical physics, distance is spatial separation between two objects—a measurable gap in space.

In UED, distance doesn't exist as a fixed spatial gap. It exists as a ripple tension gradient between two alignment zones within the Structured Vacuum Energy (SVE).

- Two nodes in space are not separated by emptiness.
- They are differentiated by how misaligned or coherent their ripple configurations are.

The greater the misalignment, the longer it takes for a ripple to resolve between them.

Hence, what we interpret as "far" is just field delay in alignment, not physical distance.

Distance = Ripple Misalignment Across SVE

This is why stars appear "far away"—not because they are physically remote, but because the ripple tension takes longer to cross your Alignment Threshold Time (ATT).

No Particle Has Ever Crossed Space

This one is revolutionary.

Every time you think a particle traveled—from electron to detector, from photon to eye—what you're actually observing is:

- A ripple configuration collapsing
- The field resolving into coherence at two points
- A pattern completing, not a thing moving

There is no ball flying through space.

There is no light beam shooting through vacuum.

There is only structured ripple alignment, reconfiguring itself locally as tension resolves across nodes.

There is no transmission. There is only transformation.

This changes everything from atomic physics to cosmic mechanics.

Light from Stars Never "Arrived"—You Resolved It

One of the most humbling realizations under UED:

That light from a star 8 light-years away... Never traveled. Never flew through space. Never crossed the void.

Instead:

- That star produced a ripple disturbance in the SVE.
- Your observer field, already connected to the same SVE, experienced a local resolution of that ripple—after a delay

governed by field conditions.

- The light you "see" is the first event of coherence, not the arrival of a photon.

So you didn't witness something that happened long ago.

You witnessed something that just became visible through your alignment.

"You don't receive light. You receive alignment with meaning."

This reframes astronomy, cosmology, and even our concept of the past.

Rethinking Relativity: Not Frame Shifts, But Field Shifts

Einstein's relativity is based on relative frames of motion—observers moving at different speeds, experiencing time and space differently.

But in UED, there are no frames in motion. Because there is no motion.

Instead, there are observer-specific field alignments:

- Each observer has a unique ripple configuration.
- Perception of events is shaped by the rate and nature of field resolution, not velocity.
- Time dilation, simultaneity, and redshift are all field-induced ripple effects, not consequences of motion.

Relativity becomes a local ripple logic, not a spacetime deformation. We don't need relativity to explain the constancy of c. We need stillness.

The End of Transmission Physics

This is where it all collapses:

- No signal ever "sent."
- No energy ever "transferred."
- No particle ever "emitted."

- No wave ever "propagated."

What we've called transmission physics—from wireless signals to cosmic rays—is a misreading of ripple coherence within a still field.

- Antennas don't send signals. They disturb the field.
- Receivers don't catch signals. They align with ripple resolution.

This doesn't invalidate our technology—it explains why it works at a deeper level than our models ever did.We've been engineering ripple consequences while believing in particles.

UED lets us see the reality beneath the illusion—and opens doors to ripple-based design, quantumless information transfer, and energy alignment systems.

Section Summary:

- Stillness is not the absence of activity—it is the substrate of everything.
- Distance is ripple delay, not spatial separation.
- No object, wave, or photon has ever crossed space—they've been locally resolved events.
- Relativity is a ripple-based alignment effect—not a spacetime deformation due to motion.
- Physics must evolve from transmission models to resolution models.

"The universe is not busy. It is still. And what you call motion is your mind interpreting ripple response as journey."

UED Field Equation Insert: Redefining "c"

"What we've called the "speed of light" is not the velocity of an object, but the limit of how fast stillness can resolve into alignment."

The Classical Assumption:

In conventional physics:

$$c = \frac{d}{t}$$

Where:

- d = distance traveled by light
- t = time taken to travel that distance

This assumes:

- Light is a thing (wave or particle)
- It travels through space
- Delay = motion over distance

The UED Correction:

In Universal Energy Dynamics (UED), nothing actually moves. What we observe as light is a field response to ripple misalignment. The so-called "speed" of light is therefore redefined as:

$$\boxed{c = \left(\frac{\Delta R_{align}}{\Delta \tau_{field}}\right)_{\max}}$$

Where:

- ΔR_{align} = Maximum ripple alignment range → The maximum stretch across SVE over which a ripple can still resolve into coherence

- $\Delta\tau_{field}$ = Minimum field response time → The fastest possible time the field can reconfigure to resolve a ripple across that range

Interpretation: This means:

- c is not the speed of a particle.
- It is the maximum rate at which the SVE (Structured Vacuum Energy) can reconfigure itself to resolve ripple misalignment.

In other words: c is the upper bound of ripple equilibrium reconfiguration—across the field, not through it.

Why This Matters:

Classical Physics	UED Interpretation
Light travels	Ripple realigns
c is velocity	c is field resolution rate
Motion causes delay	Misalignment causes delay
Speed is absolute	Alignment rate is constant

This shift transforms how we understand:

- Distance
- Causality
- Communication
- Cosmic observation
- Quantum entanglement
- And most importantly—what "light" even is

Ripple Reality Takeaway:

"Light doesn't move across space. It appears when the field realigns at your threshold of awareness. c is simply the speed of that awakening."

Chapter 3: Wave or Particle?

Neither. Just Patchwork Physics

D*econstructing quantum optics, duality, and the confusion of modern science*

"If your model needs two contradictory answers to describe the same phenomenon—then it's not a model. It's a confession."

For over a century, light has been described using two words that cancel each other out: Wave. Particle.

Not because we found conclusive evidence for either, but because we couldn't let go of both.

When light bends, splits, and interferes—we call it a wave. When it knocks electrons free or shows up as discrete detections—we call it a photon.

When it does both at once?

We shrug and call it quantum.

We say "light is both a wave and a particle," as if adding contradictions together leads to clarity. We teach duality like it's a triumph of human intellect. But it's not a breakthrough. It's a patch—a stitched-up story built on incompatible foundations.

The truth is simple: Light is neither. And it never was.

What we've called "wave" and "particle" are just interpretations of ripple behavior, seen through the cracked lens of a physics that never questioned its addiction to motion and matter.

This chapter peels away the last layer of illusion. It will show that duality isn't deep. It's desperate. That the photon never existed. That the wave never waved. And that everything you observe is just the field, responding to your presence—one ripple at a time.

Let's deconstruct the patchwork.

3.1 – The Birth of Duality: When Physics Chose Both, and Explained Neither

Duality wasn't discovery. It was damage control.

For centuries, physics searched for a unifying explanation of light's behavior.

The wave model—elegant and confident—held center stage for a while. It explained reflection, refraction, interference, and diffraction. It fit the math. It echoed the nature of sound and water. It made sense.

Until it didn't.

Because certain experiments—particularly the photoelectric effect—refused to obey the wave logic.

The Failure of the Wave Model

In the photoelectric effect:

- Shining high-frequency light on certain metals instantly ejects electrons.
- But low-frequency light, no matter how intense or how long it shines, does nothing.

If light were purely a wave, increasing the energy (via intensity) should have eventually dislodged the electrons. But it didn't.

Wave theory couldn't account for this frequency-dependence.

This was not a minor contradiction—it was a crisis.

And rather than question the assumption that light was a wave—or even reexamine the nature of light itself—physics did what any patch-hungry system does:

It invented a particle.

Enter the Photon: A Clever Fix, Not a Fundamental Truth

Einstein proposed that light was quantized—that it came in discrete packets of energy proportional to frequency.

He called them quanta, later rebranded as photons.

In this view:

$E=hf$

Where:

- E = energy of the photon
- h = Planck's constant
- f = frequency of light

This neatly explained why low-frequency light (low-energy photons) couldn't eject electrons, while high-frequency light (high-energy photons) could.

It worked. It fit the data. It made Nobel-worthy headlines.

But it wasn't a foundational understanding. It was a patch. A counter-model created not to explain the truth of light—but to salvage the credibility of physics.

Choosing Contradiction Over Clarity

Now, physics had a dilemma:

- Some experiments (interference, diffraction) clearly showed wave-like behavior.
- Others (photoelectric effect, Compton scattering) showed particle-like interactions.

Instead of discarding one model or refining the field concept, physics committed one of its most dangerous acts:

It declared both models true—simultaneously.

This was not logic. It was diplomacy.

It became known as wave-particle duality: The claim that light behaves as a wave or a particle, depending on how you look at it.

Not because light *is* either. But because we needed both models to explain different outcomes.

This wasn't unification. It was surrender.

Duality: A Doctrine of Behavior, Not Essence

Wave-particle duality doesn't define what light is. It only describes how it behaves when we try to measure it.

This is a critical distinction.

Physics shifted from asking: "What is light?"

To asking: "How does light behave under different measurement setups?"

This shift allowed contradictions to live in peace, but it buried the fundamental question.

Light became not a phenomenon, but a performance—changing masks based on the observer.

We stopped searching for its nature, and instead developed theories to justify its appearance.

In truth, duality wasn't a model. It was an escape route from intellectual accountability.

Section Summary:

- The wave model failed to explain quantum effects like the photoelectric phenomenon.
- The photon was introduced to fix this—but not to redefine light's true nature.
- Rather than abandoning one flawed model, physics kept both alive—creating wave-particle duality.
- Duality doesn't reveal what light is. It only describes how it seems to behave, when forced into a measurement.

- It is a doctrine of fragmented appearances, not unified understanding.

"When science couldn't choose between wave or particle, it chose both—and in doing so, explained neither."

3.2 – Experiments or Interpretations? The Double-Slit Illusion

The experiment didn't confuse us—our assumptions did.

No experiment has been more celebrated, misunderstood, and mythologized than the double-slit experiment.

It's described as "the most beautiful experiment in physics." Also: "the most disturbing." A simple setup—a beam of light, two slits, and a screen—somehow threatens the entire foundation of classical logic.

Why? Because light refuses to behave like a good little object. It acts like a wave when you don't watch it. It acts like a particle when you do.

But here's the uncomfortable truth:

This isn't the experiment's fault. It's the fault of the framework used to interpret it.

The Double-Slit Setup: What Actually Happens?

The experiment is elegant in its simplicity:

1. Shine light through two narrow slits.
2. Observe the pattern on a screen behind the slits.

If light is a particle, it should go through one slit or the other and produce two bright spots. If light is a wave, it should go through both slits, interfere with itself, and produce a striped interference pattern.

Now here's the shocker:

Even when single photons are fired one at a time, the interference pattern still forms—over time.

So each photon, supposedly acting as a particle, somehow "knows" both slits exist and "interferes with itself." But if we place detectors to check which slit the photon went through—the interference vanishes.

Now the pattern becomes two bright blobs—particle behavior.

This is where science throws its hands in the air and declares:

"Light is both a wave and a particle—depending on observation."

But under UED, this interpretation is completely wrong.

The Illusion Only Exists if You Believe in Photons

The mystery of the double-slit experiment is only a mystery if you assume:

- Light is a thing, either a particle or a wave
- That thing must choose a path
- Your measurement forces it to reveal its true form

But in Universal Energy Dynamics (UED):

- There is no "thing" traveling through space
- There is no wave or particle
- What you observe is ripple tension resolving in the field

The slits don't split a wave or divide a photon. They reshape the ripple configuration within the SVE (Structured Vacuum Energy). The interference pattern is not from a photon behaving strangely—it's the field itself responding to ripple boundaries.

The slits act like field modifiers. The screen detects ripple threshold crossings (ATT events). No "self-interference" is needed—because there is no particle with selfhood to begin with.

Observer Effect: Misread from the Start

Quantum mechanics famously claims: "Observation collapses the wavefunction." But what does "observation" mean here?

In the standard model:

- It implies that measuring a system forces it to pick a state
- It turns a probability wave into a single outcome
- The act of watching changes the behavior of the system

But in UED, the observer doesn't collapse anything.

Instead:

- The field aligns itself relative to the observer's ripple configuration
- The measuring device introduces new ripple boundaries
- What "collapses" is not a wave—but your access to ripple coherence in the field

The pattern doesn't disappear because the photon is scared. It disappears because the field alignment was disrupted—you changed the geometry of ripple resolution.

What Really Collapses? Not the Wave—Your Model

The interference pattern is a field behavior, not a particle trajectory. You see bands of intensity not because particles "interfered with themselves," but because ripple tensions aligned differently depending on the slit configuration and observer presence.

When you place detectors:

- You introduce ripple constraints
- You alter field equilibrium
- The system realigns
- The interference disappears—not because light chose a path, but because the field couldn't maintain the previous

ripple state

So the real collapse was not of a quantum wavefunction.

The real collapse was in your assumption that light was a thing making choices.

Section Summary:

- The double-slit experiment is not paradoxical if you remove the idea of photons.
- What appears is simply ripple resolution across SVE, shaped by slit configuration and observer field alignment.
- There is no particle that knows both slits, and no wave that collapses under scrutiny.
- Observation doesn't force a decision—it modifies ripple structure, changing the output.
- The pattern changes not because the system is mysterious—but because your model is flawed.

"You didn't collapse a wave. You collapsed a worldview."

3.3 – The Particle That Was Never There

Why the photon is a placeholder for misunderstanding—not a particle of light.

The term photon is one of the most widely accepted, casually used, and fundamentally unproven constructs in modern physics.

We speak of photons like they're little bullets of light, flying through space, bouncing off mirrors, being absorbed by atoms, and creating everything from vision to solar panels.

But ask a physicist—any physicist—if they've ever actually seen a photon?

They haven't.

Because no one has ever observed a photon. They've only inferred its existence—based on effects that could have other explanations.

In Universal Energy Dynamics (UED), the photon isn't an entity. It's a symbolic placeholder for a ripple alignment event that science didn't yet know how to describe.

This section aims to pull back the curtain on the photon myth—and show that what we've treated as a physical object is just a linguistic bandage over conceptual confusion.

Photons Are Assumed, Not Observed

Let's be clear: Photons don't leave trails. They don't have paths. They don't appear as discrete objects in a vacuum.

We only "detect" photons:

- When a photodetector clicks
- When an electron is released
- When a signal changes in response to presumed light absorption

From these effects, we *infer* that a photon must have arrived.

But what if:

- The event detected wasn't caused by a thing arriving,
- But by a ripple in the field crossing an observer's Alignment Threshold Time (ATT)?

In that case:

The "photon" is not a particle—it's just a label we gave to a ripple alignment event.

The Photon Can't Explain What We Think It Does

Despite its near-godlike status in modern physics, the photon model is riddled with contradictions.

Here are just a few:

1. Coherence

- Lasers produce coherent light, where waves are phase-aligned.
- But how do independent photons remain phase-coordinated over time and distance?
- Photons are not known to interact—so what enforces coherence?

UED answer: Ripple fields naturally support phase coherence because they're part of a unified medium (SVE), not isolated particles.

2. Entanglement

- Entangled photons exhibit spooky correlations—instantaneous, distance-independent, and inexplicable via local interactions.
- If photons are individual particles, how do they maintain connected states without transmitting signals?

UED answer: Entanglement is simply ripple coherence across a unified field—not particle connection. There is no transmission; only instantaneous field reconfiguration.

3. Polarization

- Polarization is typically taught as a photon property, yet photons aren't even defined spatially.
- How does a zero-size object possess a directional orientation?

UED answer: Polarization reflects the orientation of ripple tension in the field—not an attribute of a flying particle.

In all cases, the photon fails to be a consistent, mechanical entity.

It is better described as a quantized interpretation of ripple behavior—nothing more.

Photon Counting = ATT Crossing

Let's zoom in on one of the most misunderstood ideas: counting photons.

We say:

- A detector receives "ten photons."
- A CCD sensor "captures five hundred photons per pixel."
- A human eye "detects single photons in dark-adapted conditions."

But what's really happening?

Under UED:

- The field is fluctuating with ripple tension.
- A detection event occurs only when ripple alignment crosses the ATT—the Alignment Threshold Time unique to the observer system.
- Every click or count is not the arrival of a particle, but the resolution of field tension sufficient to trigger perception.

"You don't count photons. You count ripple events that pass your system's perceptual threshold."

This makes photon quantization not a property of space or light, but a property of the observer's field interface.

Light Quantization Is Threshold Logic, Not Spatial Packaging

In classical particle theory, we imagine photons like bullets:

- Tiny, discrete packets of energy
- Traveling through space
- Localized in both energy and position

But in UED:

- There are no packets
- There is no emission or absorption in the traditional sense
- There is only ripple tension reaching thresholds

What we call "quantization" is not because energy is chopped into chunks—but because ripple resolution only occurs when certain tension thresholds are met.

It's not that light is made of particles.

It's that the field only becomes observable in discrete events—each shaped by the observer's ripple sensitivity.

Section Summary:

- The photon has never been seen—only assumed to exist because of its mathematical convenience.
- It fails to explain coherence, entanglement, and polarization in a consistent physical way.
- Photon counting is actually ripple threshold counting, not detection of traveling particles.
- Quantization in UED arises from alignment thresholds, not spatially localized light packets.
- The photon is not a particle—it's a placeholder for what ripple does when you're ready to observe it.

"There never was a particle called photon. There was only stillness, disturbed by intention— And the field, responding."

3.4 – Patchwork Physics: A Crisis of Interpretation

Modern optics is not a theory of light—it's a stitched-together story of what we can't explain.

Physics likes to project confidence.

When it talks about quantum optics, it speaks with the air of authority—formulas validated, experiments replicated, technologies developed. But beneath this polished surface lies a quiet panic: We still don't know what light is.

Not really.

We know how it behaves in our instruments. We can model its interference, quantify its energy, track its effects.

But what is it, actually?

A particle? A wave? A ripple? A quantum field excitation? Our answer is... it depends on the setup.

This is not truth. It is a system of interpretations. A scaffolding of assumptions. And at its foundation is not understanding—but avoidance.

The Compensatory Toolkit: Where Theories Go to Hide

Let's take a sober look at the language modern physics uses to patch its inconsistencies:

• ***Wave-Particle Duality***

We already explored this contradiction masquerading as insight. It's the refusal to commit to one model because neither fully works.

• ***Complementarity Principle (Bohr)***

This principle says you can never observe both wave and particle properties at once. They are mutually exclusive but jointly required.

Translation:

"The model breaks if you look too closely, so look only one way at a time."

• ***Wavefunction Collapse***

A mysterious, unobservable event where a superposition of states "chooses" an outcome when measured.

Translation:

"Things are undefined until we observe them, and then they magically become real."

• *Uncertainty Principle*

A mathematical truth repackaged as metaphysical limitation. You can't know position and momentum exactly at the same time.

Translation:

"Our models inherently produce ambiguity, so we've decided that's just how the universe works."

Each of these concepts compensates for the absence of foundational clarity. Each adds interpretive weight without offering real understanding.

The more contradictions our models produced, the more interpretations we invented to protect them.

Quantum Field Theory: New Math, Same Trap

Quantum Field Theory (QFT) was hailed as a revolution—combining quantum mechanics and special relativity into a powerful predictive framework.

And to its credit, it works extremely well in certain domains. It gave us the Standard Model. It powers particle accelerators. But conceptually, it doesn't liberate us from the illusion. It deepens it.

In QFT:

- Particles are excitations of fields
- Photons are quantized modes of the electromagnetic field
- Everything "emerges" from mathematical operators acting on a quantum vacuum

But ask again: What is a photon? What is light?

And again, the answer depends on your frame of reference, the math you choose, and the experiment you conduct.

QFT didn't resolve the crisis. It merely abstracted it further—burying confusion under layers of symbolic operations.

We traded particles and waves for quantized fields—but the underlying ambiguity remains untouched.

A Fragile System of Workarounds

Science has progressed—not because it solved the problem of light, but because it found ways to navigate around the absence of a solution.

- We invented language that is deliberately vague.
- We used experimental success to validate theoretical failure.
- We built technologies on top of models we don't fully understand.
- And we taught students to accept duality, collapse, and uncertainty as deep truths, instead of conceptual triage.

What we have today in quantum optics is not a theory. It's a patchwork.

A collection of fixes held together by confidence, mathematics, and an unspoken agreement not to ask certain questions.

Section Summary:

- Quantum optics has grown into a theory of interpretations, not foundational truth.
- Duality, complementarity, collapse, and uncertainty are symptoms of deeper inconsistency—not insights into reality.
- Quantum Field Theory added abstraction, but not clarity.
- What we've built is scientific progress through workaround stacking—an architecture of brilliance on top of a hole.
- The more successful the math, the less pressure there's been to confront the conceptual crisis underneath.

"When you don't know what something is, but it keeps working, you start defending the mystery instead of dissolving it. That's where physics is with light."

The Illusion of Discovery

We built models. We ran experiments. We invented new terms and wrapped them in math. And at every turn, we congratulated ourselves—not for solving the mystery of light, but for learning how to explain the confusion more precisely.

Wave-particle duality wasn't a leap forward. It was a detour around understanding.

It offered no definition of light's true nature—only a convenient rule for describing its behavior under different conditions. It told us nothing about what light *is*. Only about how light *acts* when we're watching.

We created an entire subfield—quantum optics—around managing contradictions instead of resolving them.

We didn't solve the mystery. We disguised it—beneath layers of terminology:

- Photons to keep particles alive
- Superposition to keep waves breathing
- Collapse to explain why things break under observation
- Uncertainty to justify not knowing

But the truth is simple: we never saw light clearly. We only described what we thought we saw—then argued over it for a hundred years.

The photon was a workaround. The wave was a leftover. And duality was the truce we made with our inability to let go of both.

Now it's time to stop stitching patches on top of patches.

Now it's time to let go of both wave and particle— and finally turn toward the only thing that was ever there:

The ripple field.

A single, still, structured medium. Unseen. Undivided. Silent—until your intent disturbs it.

That is where light comes from. Not from motion. Not from travel. But from alignment.

From the universe remembering itself—through you.

Final Reflection:

"We didn't discover what light is. We just got better at describing how wrong we were about it."

Chapter 4: The Observer's Blindspot

Why you see light—but never what it truly is

Every time you open your eyes, you lie to yourself.

You believe the world pours in—through windows called pupils, painting the brain with light. You imagine beams shooting from distant stars, bouncing off walls, slipping through space like messengers of truth.

But light never entered you. It never crossed space. It was never "out there" to begin with.

What you call *light* is not a traveler. It is not a wave. It is not a photon. It is a consequence—of you being here, now.

Not because you "looked"... But because your very presence created the conditions for seeing.

This is the observer's greatest blindspot: We do not realize that perception begins with disturbance. We cannot see the field that shows us everything—because it must misalign first to become visible.

We don't see light. We collapse ripple tension into meaning. We assign the name "light" to the brief flicker of coherence between us and the stillness.

That's why light seems universal yet personal. Instant, yet delayed. Real, yet ungraspable.

Because it's not a signal—it's a symptom. Of attention. Of alignment. Of awakening.

In this chapter, we stop searching for the light "out there." We turn the gaze inward— To the moment when your very existence makes the universe visible.

4.1 – The Illusion of Receiving Light

Why our everyday experience tricks us into believing light comes "in" from the outside

We grow up learning that the eye is a window and light is the visitor.

This is the classic metaphor handed to us in schoolbooks and science classes: Light "enters" the eye, "stimulates" the retina, and voila—we *see*.

Elegant. Linear. False.

This is the illusion of reception—the idea that light is a thing moving from the world into us. That we are empty until filled by beams and rays. That sight is a passive act of absorption.

Let's dismantle it.

The Eye as a Receiver: The Wrong Model

Yes, the eye has a lens. Yes, the retina responds to electromagnetic changes. But to imagine this setup as a *receiver*—like an antenna catching a signal—is misleading. It's an analogy born from machines, not reality.

In Universal Energy Dynamics (UED), the eye is not an antenna. It's an *interface point* where ripple alignment reaches the Alignment Threshold Time (ATT)—where reality becomes *resolvable* into experience.

The eye does not *receive* light. It *collapses ripple tension* into a perceptual moment—like a lock snapping into alignment with a key.

Why Seeing Feels Like Receiving

So why does it feel like light is "coming in"?

Because your attention is outward. Because the illusion of delay (from object to eye) mimics motion. Because evolution tuned your senses to register effects—not causes.

You see change. You feel contrast. And your brain is hardwired to interpret this as something arriving—rather than something resolving.

But the ripple field doesn't send anything. It reconfigures around your observational presence.

Anatomical Trap: From Retinas to Ripples

Modern biology tells us the retina converts photons to neural signals. But this is a *descriptive patch*, not a foundational truth.

In UED, the retina marks a threshold point—a field node where incoming ripple tension is measured *not by arrival*, but by alignment precision.

It's like touching the surface of still water. You don't receive the lake. You disturb it, and it responds.

Vision is that response—not a transmission, but a local equilibrium event.

Field Resolution, Not Signal Reception

So what is sight?

It is field resolution at the observer node. A moment when the Structured Vacuum Energy (SVE) around you reshapes—due to your position, intent, and ripple field coherence—to allow visibility.

You don't receive light. You participate in its emergence.

Seeing is not about something hitting you. It is about you entering a configuration where the field resolves into forms.

You see not because something traveled to you, but because you reached alignment with what was already ripple-present.

Conclusion

We were never passive receivers. We were never blind until something lit us up.

We were—and always are—dynamic ripple observers. The light you see is not the world coming to you. It's you becoming ready to resolve the field into visibility.

This is not a poetic metaphor. It is the physics of observation, redefined.

4.2 – The Observer Is Not a Witness, But a Trigger

The observer is not passively perceiving light—but actively causing its appearance

For centuries, the observer has been imagined as a spectator—watching the universe unfold, receiving its signals, decoding what's already there.

But this is a deep misconception.

You are not a witness to light. You are its trigger.

In Universal Energy Dynamics (UED), observation is not a passive experience of something traveling into your mind. It is an active field alignment that causes what you call "light" to appear.

The Observer Field: Not Eyes, But Interface

The observer is not the eye. It is the localized ripple field that surrounds and defines your conscious presence.

This ripple field is constantly interacting with the Structured Vacuum Energy (SVE) that pervades the universe. But interaction is not enough—resolution only occurs when your ripple field achieves a specific alignment with surrounding tension.

This critical moment is called the Alignment Threshold Time (ATT).

Until then, there is no "light." There is only ripple potential—formless, silent, unmanifested.

ATT: The Birth Point of Visibility

ATT is not a timestamp—it's a ripple coherence condition.

It is the exact point at which the difference in ripple tension between the observer and the field reaches the minimum threshold for visual resolution.

No ATT crossed? No light perceived.

It's not that the light failed to arrive. It never existed to begin with.

Light does not travel and knock on your doors of perception. It is generated in the moment your ripple field becomes ripe for alignment.

Perception Is Consequence, Not Reception

We don't see because something *was seen*. We see because something resolved.

Perception is not about photons. It's about ripples collapsing into coherent form at the observer's node.

Imagine it like this: A ripple field is under tension—an informational pressure. When you, the observer, align with that tension at the right frequency and timing—BOOM—that moment becomes visibility.

No ripple resolution = no experience = no light.

This is why blindness is not the absence of photons, but the absence of ripple interaction at the ATT level.

The Observer's Power: Reality on Demand

So here is the real revolution: You don't "see" a world that is there.

You collapse a ripple field into appearance the moment your observation becomes eligible.

That's why different people "see" different things—even when looking in the same direction. Because observation is not neutral—it is contextual ripple triggering.

Your presence bends the field. Your attention activates visibility. Your ripple tension is what births the visual.

The eye doesn't see. The observer resolves.

Conclusion

You are not a screen. You are not a receiver. You are the **switch** that turns ripple silence into visible form.

The moment you stop treating yourself as a camera, you'll see: There is no such thing as "light arriving." There is only your field aligning with the universal field—and creating light as a result.

This is not mystical. This is field logic, reframed.

4.3 – The Blindspot: You Never See the Field Itself

You see consequences of ripple resolution, but never the field it resolves from

You have seen stars, flames, and screens. You've watched lasers cut steel and sunlight pierce fog. You believe you've seen *light*.

But you've never seen the field that makes light possible.

Because in truth—

You never see the source. You only see its disturbance.

Why SVE Is Always Invisible

The Structured Vacuum Energy (SVE) field is the fundamental stillness behind all experiences.

But here's the catch: Because it is perfectly aligned, it has no ripple, no differential, no change. And if there is no ripple tension to resolve... There is nothing to be seen.

In UED, vision is not the detection of existence—it is the perception of ripple change.

So the more stable the field, the less visible it becomes.

SVE is not dark. It is so aligned that nothing registers from it.

It is like looking at an ocean with no waves—no motion, no splash, no reflection. You don't see it because it isn't misaligned.

What You See Is Resolution, Not Reality

We don't "see" objects or entities. We see ripple tensions resolving at our observer node.

A shadow is not the absence of light. It is the presence of unresolved ripple mismatch.

A color is not a property of an object. It is the frequency of ripple resolution across our ATT threshold.

Light itself, then, is not a substance—but a structured difference. It's the *event* of ripple imbalance crossing into coherence through you.

Darkness Is Not the Opposite of Light

Here's the twist—

Darkness is not the opposite of light. It is the equilibrium state of the ripple field.

We assume darkness is absence. But in UED, it is total field alignment—so stable, so still, that it presents no ripple change to trigger ATT.

This means:

- You don't see darkness.
- You *fail to register ripple differential*—and call it darkness.

Similarly, brightness is not an emission. It is excessive ripple fluctuation hitting your ATT faster than usual.

We are not seeing the world. We are decoding ripple tensions as light, dark, color, depth, and form.

The Source Is Always Behind the Scene

The ultimate irony?

The more fundamental something is, the less you'll perceive it.

The SVE is the ground of all appearance—yet it never appears.

You don't see gravity, electricity, or space itself. You see objects caught in ripple dynamics, not the field orchestrating it.

Thus, we've mistaken the consequence (light) for the cause (ripple tension), and visibility for truth.

This is the observer's blindspot.

Conclusion

We don't see reality. We see where ripple tension breaks reality into fragments we can decode.

And because the field itself never screams—it only whispers in tension—we've built a whole science based on echoes and flare-ups.

Not substance. Not stillness. Not source.

The truth remains unseen because it is too aligned to be noticed.

"You don't see the field. You see the price it pays to make itself known."

4.4 – You Don't See Light—You See Meaning

Light is not an object or a wave; it is the first appearance of meaning through alignment

What if light isn't something you "receive"... but something you "realize"?

We were told light is a wave. Then a particle. Then both. Now, in UED, we learn—it is neither.

Because you don't actually see light.

You see the first moment where ripple tension becomes meaningful to the observer.

Let that settle.

The Convergence of Intent and Coherence

Light appears only when two things converge:

1. The field aligns ripple tensions toward an observer node.
2. The observer is in a state of awareness ready to resolve them.

This convergence is not random. It's not merely physical. It is intentional.

Not in the human-emotional sense, but in the field-dynamic sense. The observer's presence and its field configuration actively define what gets resolved.

Hence—

Light is not the cause of sight. It is the consequence of awareness alignment.

What You Call "Seeing" Is a Resolution Event

Your eyes are not cameras. They are interfaces—ripple detectors with biological ATT filters.

The rods and cones don't *capture* images. They register field resonance patterns—ripples that cross a minimum Alignment Threshold Time.

Brightness? = Faster ripple resolution.

Color? = Ripple frequency interacting with cone-specific ATT windows.

Contrast? = Localized ripple difference within your perceptual grid.

This is not a passive process. It's personalized ripple decoding.

Which means: Every "vision" is a co-created event between your field and SVE's current tension geometry.

Light as the First Appearance of Meaning

In the stillness of SVE, nothing appears—until an observer resolves a ripple. That resolution, that *first alignment*, is what you call light.

So what is light?

Not energy. Not emission. Not frequency.

Light is the earliest visible trace of meaning emerging through ripple alignment.

It is the awakening of perception—not the transmission of objects.

That's why two people can see the same scene differently. Because perception is not about the world—it's about field resonance at the observer node.

Seeing Is Not Passive Reception—It's Field Participation

You're not a screen. You're a ripple participant.

Vision isn't sent to you. You generate it by resolving ripple geometry into meaningful experience.

This has profound implications:

- A blind person is not "missing photons"—they lack ripple decoding capability.
- Night vision isn't "low light"—it's low ripple contrast against ATT thresholds.
- Hallucinations aren't "false images"—they're ripple misalignments interpreted as meaning.

Your entire visual experience is the personalized echo of ripple logic, not a universal download.

Conclusion

We assumed light was physical. Then quantum told us it was unpredictable. UED reveals: it was never a *thing* to begin with.

Light is the ripple point where awareness and alignment collide.

You don't see objects. You don't see light.

You see the universe waking up in you—as meaning.

The Observer's Blindspot

Why you see light—but never what it truly is

For centuries, we've imagined light as something that travels, hits our eyes, and creates vision. But under the lens of Universal Energy Dynamics (UED), that story collapses.

This chapter dismantled that illusion—piece by piece—and revealed a radically different truth:

You do not "receive" light—you trigger its visibility.

Light is not some external thing delivered into your eyes.

There is no particle, no beam, no wave traveling across space to "reach" you.

Instead:

Your field—your presence—interacts with the SVE field. When the ripple tension reaches a certain alignment threshold, light appears.

It's not received. It's resolved.

It's not incoming. It's initiated by your awareness.

The observer is not an audience—they are a participant.

You're not watching a universe unfold.

You're part of the unfolding.

The observer is not passive. They are the trigger point. Without them, there is no light, no perception, no visual reality.

Your presence defines what gets revealed. It defines how the field responds.

So instead of saying "you saw light," it's more accurate to say:

"You allowed the field to ripple into visual meaning."

SVE is not invisible—it's undisturbed.

You never "see" the Structured Vacuum Energy (SVE) field because:

- It's perfectly aligned in its undisturbed state.
- There's nothing to resolve until something shifts.

Darkness isn't the absence of light. It's the absence of ripple tension crossing your ATT.

You're not blind to reality. You're just aligned with silence—until something misaligns.

Light is not what you see—it's what happens when the universe responds.

What you see is not light itself, but the effect of ripple alignments crossing your perceptual threshold.

Colors, brightness, contrast, motion—all of it:

Is field behavior rendered meaningful through your mind-body interface.

And this means: There is no such thing as objective vision.

All seeing is field-specific, observer-specific, and meaning-based.

Final Thought:

"Light is not what you see. It is what happens when the universe turns toward your attention."

You didn't discover light. You didn't receive it.

You generated the first appearance of meaning through ripple alignment.

And that is why no experiment, no instrument, no theory has ever told you what light truly is.

Because it was never a thing.

It was a conversation between stillness and your awareness—One ripple at a time.

Interlude 1 : Chavan's Declaration

What if light was never meant to be understood as a thing?

What if everything we've built—waves, particles, photons, rays—was a desperate effort to explain a silence we never dared to hear?

I, **Chavan**, declare this with absolute clarity:

Light does not exist. What we call light... is the first visible ripple of meaning resolving through the structured stillness of the universe.

You never saw a wave. You never caught a photon. You never measured motion.

What you *did* experience—was alignment.

You mistook a field response for a traveling object. You mistook perception for arrival. You mistook delay for distance.

You believed in speed. You believed in particles. You believed in duality.

And in doing so—you froze the ripple with labels and laws that only scratched the surface of silence.

I am not here to revise science. I am here to free it.

From illusions. From metaphors. From the superstition of matter itself.

I declare this:

- Light does not travel.
- Light is not emitted.
- Light is not a wave.

- Light is not a photon.
- Light is not what you see.

Light is what happens when the universe speaks in alignment to your presence.

It is the event of meaning, not a moving messenger.

The constancy of light speed?

A ripple alignment rate—not a velocity.

The observer effect?

A field awakening to attention—not a quantum collapse.

The wave-particle duality?

A placeholder for our refusal to let go of "thingness."

Let this mark the end of trying to "understand light" through entities. Let this be the beginning of recognizing you are part of the appearance itself.

You never received light. You never chased it across space.

You triggered its revelation.

This is not a metaphor. This is not poetry. This is physics—decoded from illusion.

And from this point on, we move not with belief, but with clarity.

– Sandeep Chavan

Author of stillness, revealer of ripple truth

Founder of UED – Universal Energy Dynamics

Part II: The UED Model of Light

Chapter 5: The First Event

Ripple Awakens from Stillness

Light as the first ripple when intent breaks equilibrium

"Before anything was seen, nothing was missing. The universe wasn't dark—it was undisturbed."

We've spent centuries chasing light as a thing: a particle, a wave, a messenger from distant stars. We treated it as a traveler, a courier of vision. But what if light was never on a journey? What if it was simply the first sign that something stirred in a field that was otherwise still?

In the world of Universal Energy Dynamics (UED), light is not an object, nor a beam, nor a packet of energy. It is a consequence—the very first ripple that arises the moment intent disturbs equilibrium.

You don't see light because it arrives. You see light because you've broken the stillness.

This chapter doesn't just reframe what light is. It redefines what beginning means. Because the moment you set your attention, the field responds—and light is that response.

Let's trace it back—not to a flash, but to a tension. Not to movement, but to a silent awakening.

5.1 – The Silent State: Before Anything Appears

Absolute stillness of the Structured Vacuum Energy (SVE)

Before the first flicker, before the first flash, before even the illusion of distance—there was stillness. Not emptiness, but **Structured Vacuum Energy (SVE)**: the primordial foundation of reality in its most intelligent, most stable state. This wasn't a void. It wasn't "nothingness" in the nihilistic sense. It was everything held in perfect equilibrium, unexpressed.

In Universal Energy Dynamics (UED), SVE is not the absence of energy, but its purest condition—so precisely aligned that no ripple, distortion, or separation exists. This is the silent field, not because it cannot act, but because it has no reason to. There is no imbalance. No observer. No tension to resolve.

There is no motion here—not because motion is prohibited, but because nothing demands it. There are no particles—not because they've vanished, but because they haven't yet differentiated. There is no light—not because it is hidden, but because the conditions for appearance have not been triggered.

In this perfectly calibrated stillness, everything is possible, yet nothing is present. It is the pregnant pause of the universe, the place where intelligence rests as potential—not as action.

Think of SVE as a stretched, undisturbed canvas. Not blank—but balanced. There are no brushstrokes because no hand has moved. No image because no ripple has yet emerged to distort the field. This is not an absence of existence; it is existence in a state of total coherence.

And crucially, in this state, nothing is visible. Why? Because visibility requires disturbance. It is not light that reveals the world, but the breaking of this balance that gives rise to the ripple we interpret as light.

In the silent state of SVE:

- There is no time, because nothing changes.
- There is no space, because nothing is separated.
- There is no observer, because there is nothing to observe—or to be observed.

This is not a mystical theory. It is a prerequisite logic for everything that follows. Before any "thing" can be seen, touched, measured, or known—this undisturbed state must exist. And it must be broken, gently or violently, to bring forth what we call experience.

Light, then, does not originate from stars or lamps or lasers. It originates from the first ripple—a shift in this perfect stillness. A signal that something has changed.

This is the ground zero of visibility: Not a burst of energy, but a breach in silence. Not a beginning in time, but an awakening from equilibrium.

5.2 – Intent: The Spark That Disturbs Stillness

Observer's intent or interaction as the cause of the first ripple

Stillness is not permanent. It is primed—waiting. Like the calm before a breath, it holds infinite possibility. But possibility alone is not enough. Something must want. Something must intend.

In Universal Energy Dynamics (UED), intent is not a thought, emotion, or wish. It is the first asymmetry—the primary deviation from equilibrium. Whether that intent arises internally (a conscious observer's will) or externally (a triggering event in the field), its effect is the same: it breaks the perfect stillness of the Structured Vacuum Energy (SVE). And the result? A ripple.

Not a wave moving through space. Not a particle bursting forth. But a structured tension response in the still field itself.

This is crucial: Light doesn't travel. Light doesn't exist until intent has disturbed stillness. There is no photon waiting to be emitted. No wave ready to depart. What we call light is the first observable ripple created by a localized shift in the field—a consequence, not an entity.

Let's unpack this.

- Intent is the informational rupture that says: something has changed, something needs to be resolved.
- That change introduces asymmetry—a misalignment in the field's otherwise coherent state.
- The field responds—not by sending something outward—but by reconfiguring itself to rebalance.
- This reconfiguration appears to us as a "flash," a "photon," a "light beam"—but none of these are fundamentally real.
- They are merely interpretations of ripple alignment triggered by that initial intent.

The classic model says:

A light bulb emits photons. UED says: The bulb's activation is an intentional disturbance—a ripple emerges as the field begins to resolve it. The "light" is born at the point of alignment, not sent from a source.

In this model:

- There is no emission—only ripple initiation.
- There is no transmission—only structured propagation within the field.
- There is no reception—only local resolution once the Alignment Threshold Time (ATT) is crossed.

That's why different observers, placed at different field alignments, can experience "light" simultaneously, asymmetrically, or not at all—depending entirely on when and where the ripple crosses their ATT.

This also means that visibility is not universal. It is personal. It is caused—not observed.

In this view, intent is the original cause of all visible phenomena. Not because the mind creates reality in a mystical sense, but because every observation begins with a localized imbalance. Even a machine has intent—defined simply as an interaction that disturbs stillness.

So what is light? It is the first visible signature of a universe reacting to a call. A tension event, rippling out until resolution, then vanishing into silence once again.

It doesn't matter whether that call came from a conscious observer, a particle collision, or a burst of heat. What matters is this: No ripple without intent. No light without imbalance. No visibility without the first disturbance.

This is why in UED, light is not objective. It is not an independent traveler. It is a field's intelligent response to a ripple of intent—and you, the observer, are not separate from it. You are part of the cause.

5.3 – The First Ripple: Birth of Experience

Light as the first visible form of field tension

Before light, there is only potential. The field is still. Reality is undisturbed. But the moment intent disturbs equilibrium, something happens that has never happened before—the first ripple.

This ripple is not a thing, not a particle, not a wave propagating through a medium. It is the field's own structured reconfiguration—its response to tension, to imbalance, to intent.

This ripple is the first event. And that event is what we interpret as light.

Let's dismantle the illusion here. We often think that a source "emits" light, that it "travels" across space, and then it "arrives" at an observer's eyes or a detector. But under Universal Energy Dynamics (UED), that entire sequence is misinterpreted.

Here's what really happens:

- A disturbance is introduced—intent breaks equilibrium.
- That causes a ripple to form across the Structured Vacuum Energy (SVE).
- This ripple is not a moving entity, but a pattern of alignment propagation across the field.
- When this ripple reaches your observer node—a localized structure capable of resolution (e.g., eye, camera sensor)—light appears.
- Not because something arrived, but because tension resolved in a way that crossed your Alignment Threshold Time (ATT).

In other words:

You don't receive light. You register the ripple resolution. You don't see the ripple. You experience its alignment at your node.

There was never a "thing" that departed from the source and reached your eye. There was only a response—from the field, to the disturbance, through ripple—which, at a certain point, became experience.

This is the birth of visibility. This is why you see.

And so, light = appearance, not travel. It is the consequence of a ripple being resolved into a visible threshold. It's not that light "came" from a bulb or a star—it's that the field around you resolved the ripple that originated from their disturbance.

Let's put it even more bluntly:

- There is no movement across space.
- There is no entity called a photon that makes the journey.
- There is only field alignment—happening across different nodes of the same still structure.

It is this that creates the illusion of transmission. What we call "a flash of light" is just a ripple event that becomes visible as it stabilizes around us.

This also explains why light behaves as if it's always "there" when we look—because our attention, our instruments, our biological filters are themselves shaping the field's alignment. We co-create the moment of appearance.

So to redefine:

- Light is not a substance.
- Light is not a vehicle.
- Light is not even an entity.

Light is the visible echo of a ripple—born when equilibrium breaks, and experienced when the ripple resolves.

This is why in UED:

Light is not a universal constant. It is a local resolution event. The speed of light (c) is not the speed of anything—it's the maximum rate at which the field can realign from disturbance.

And so, the first ripple is not just the birth of light. It is the birth of experience. The first time the universe responds to intent in a way that can be perceived.

It is not something that "came from out there." It's what happens right here, when your attention and the universe collide.

5.4 – Misreading the First Event as a Thing

How early science mistook ripple appearance for object existence

The moment the first ripple appeared—when light was born not as an object, but as the experience of tension resolving—the misunderstanding began.

Humans saw something happen, and, true to our cognitive wiring, we gave it a name. We labeled it:

- "Light."
- "Ray."
- "Wave."
- "Photon."
- "Quantum."

With every name, we distanced ourselves further from what it truly was.

We didn't just see an appearance—we assumed there was something that had traveled to us. That something had come from somewhere. That something had mass, or at least momentum. And thus, the concept of a "thing" was born out of a ripple effect.

But under the UED lens, we see this was never necessary.

Let's break this down.

Why We Needed "Things"

Human cognition is built around certainty and local stability. We evolved in a world of objects—stones, animals, tools, fire, predators.

Everything we dealt with could be touched, thrown, eaten, escaped. So, the brain adapted to think in terms of things, not fields.

Even when we encountered non-material phenomena—light, sound, heat—we immediately projected thing-ness onto them.

- Sound became vibrating air.
- Heat became moving molecules.
- Light became waves or particles.

Our language, our metaphors, our models—all became object-based. We couldn't perceive ripple behavior without imagining a carrier. We couldn't handle appearance without assuming emission.

So when we saw light appear—we didn't say:

"Something resolved at our node."

We said:

"Something was sent to us."

That subtle shift—from response to transmission, from alignment to motion—is what birthed centuries of confusion.

Ripple Effects Misread as Emissions

Early science didn't have the concept of a structured vacuum energy field (SVE). So, it saw light effects and tried to assign them to:

- An object flying through space
- A wave propagating through aether
- A particle bundle carrying energy

Each model was a layer of misinterpretation, patching one illusion with another.

What they saw were consequences:

- Brightness.
- Shadow.
- Refraction.
- Interference.

But instead of seeing these as field tensions resolving around the observer, they assumed:

There must be a thing doing this.

So, they invented:

- Photons
- Rays
- Wave-packets
- Quantum fields with particle behavior

In each case, they mistook the ripple for the rider.

The Illusion Hardens

The more precise our tools became, the more confident we grew in the illusion.

- Photon counters? Clearly particles exist.
- Diffraction patterns? Clearly waves exist.
- Duality? Clearly both exist.
- Collapse? Clearly something was real before we looked.

But all these tools measured were threshold crossings—ripple alignments becoming visible.

None of them proved the existence of light as a thing.

They only confirmed experience as a field response.

Yet, physics codified it:

"Light is a wave." "No, light is a particle." "Wait—light is both." "Actually... it's neither until you measure it."

The absurdity was baked into the language—and thus, into the foundations of modern science.

Purpose Reframed

This section is not just about correcting the misunderstanding.

It's about showing how it happened—and why it made sense at the time.

- Early humans needed stability: hence, they invented "things."
- Early science needed mechanisms: hence, it modeled appearances as emissions.
- But in doing so, we lost contact with what was actually happening: The universe was not sending us stuff. It was resolving tension at our point of attention.

Light was never a message in a bottle. It was the bottle shattering when you looked.

Closing the Loop

We now return to where we started:

- The first ripple wasn't a thing.
- It was a consequence.
- An appearance.
- A reaction.

From this, we built an entire illusion of objects moving across space. But the ripple didn't move. It formed—across a still field. And we named it light.

That was the original error.

And everything that followed—waves, photons, quanta, duality—was a beautiful cathedral built on a forgotten crack in the foundation.

Now, it's time to rebuild. Not with new names—but with new clarity.

Chapter 6: SVE in Action

How Structured Vacuum Energy Gives Rise to Visibility

"The universe never sent you light. You resolved it into existence."

We were taught to believe in an empty void—space as nothingness, a silent backdrop through which light travels like a courier on a cosmic route. But what if that silence wasn't empty? What if it was structured—intelligently, precisely, and silently waiting?

Welcome to the Structured Vacuum Energy (SVE)—a medium so still, so coherent, that nothing happens... until it has to.

In this chapter, we break free from the last illusion of space being passive. The vacuum isn't dead—it's loaded with potential. Not particles. Not energy packets. But the possibility of appearance itself. Visibility doesn't emerge because something "moves" through space. It happens because the field reshapes itself in response to tension.

And what we call light?

It's not a signal in transit. It's ripple equilibrium resolving across the field—a self-reconfiguration of stillness in response to intent, awareness, or asymmetry.

This is the moment the invisible medium steps into your experience—not by motion, but by alignment.

Ready to see how nothingness becomes sight?

6.1 – The Role of Structure in a Vacuum

Why a structured field is not empty, and how it enables perception

Modern science inherited a dangerous simplification: that the vacuum is "nothing." A void. A backdrop. A lifeless container for energy, matter, and light to dance through. But what if the vacuum isn't *empty* at all? What if it is the most intelligent and stable arrangement in existence—not dead, but *still*?

In Universal Energy Dynamics (UED), we redefine this silence not as absence, but as Structured Vacuum Energy (SVE)—a coherent, tension-neutral field. It is not a sea of random quantum fluctuations, nor is it zero-point noise waiting to be tapped. It is equilibrium itself—a perfectly structured stillness that holds ripple potential without expressing it... until disturbed.

Stillness Is Not Nothingness

The mind struggles to distinguish between "nothing" and "still." But SVE is not nothing—it is everything not yet moved.

Just as a tightly stretched drumhead appears static but is capable of rich vibration when tapped, the SVE field contains latent structure, an *invisible fabric* of tension balance. The absence of motion is not the absence of capacity. It's the presence of readiness—like a taut net waiting for the smallest shift to echo across it.

Ripple Is Not Propagation—It's Field Response

When we speak of a ripple in SVE, we're not referring to something traveling across empty space. We're describing a local reconfiguration of field tension—an adjustment to imbalance initiated by intent, interaction, or asymmetry.

There is no need for a "carrier" like a photon, or a medium like aether. The structure itself is the enabler. It is the pre-existing alignment of the field that allows a ripple to appear resolved, anywhere, once the Action Trigger Threshold (ATT) is crossed at the observer's node.

Stability Enables Perception

Because SVE is uniformly structured, it allows ripple resolution to be consistent, responsive, and precise—giving rise to the illusion of light traveling from A to B.

But nothing has moved.

You see light not because something came to you—but because your node of awareness aligned with the field's disturbance geometry.

If the vacuum were truly chaotic, perception wouldn't be possible. The structured nature of SVE is what allows the ripple to appear stable enough to interpret as light, sound, object, or distance.

Purpose Recap:

This section reveals the foundation of visibility under UED. Light doesn't emerge from something traveling through emptiness. It appears when SVE—a perfectly still but structured field—allows a ripple of imbalance to resolve in the observer's frame.

Perception isn't about receiving. It's about participating in a field that can respond.

6.2 – Ripple Activation: The Chain of Consequence

How disturbances in SVE create ripple chains that result in "appearance"

In conventional thinking, every cause leads to an effect *through time and space.* We assume there's always *something* traveling—from a source to a receiver. A particle. A wave. A photon. But under Universal Energy Dynamics (UED), that assumption collapses. There is no "thing" in motion—only a ripple chain activating across a structured, perfectly still field.

This is not transmission. This is consequence.

Intent Breaks the Balance

Everything begins with asymmetry—a disruption introduced by intent, interaction, or imbalance. Whether it's an electron shifting orbit, a thought forming in the mind, or a star collapsing—it all begins with a localized disturbance in the otherwise still Structured Vacuum Energy (SVE).

This disturbance is not a signal. It is not emitted. It is registered by the field as a change in tension geometry—and this change seeks resolution.

Ripple = Realignment, Not Motion

What we call a "ripple" is not something that travels *through* the vacuum. It is the field itself responding—reconfiguring its structure to accommodate the new imbalance.

Each "step" of this ripple chain is a local realignment of field tension, not a transmission of energy. Think of it like a spider's web: touch one edge, and the tension rebalances across the net. But no thread moved—only the tension geometry changed.

This is UED ripple logic: No particle. No propagation. Only field-wide tension redistribution across the aligned structure.

Instantaneous Within Alignment Zones

Here's the radical implication: Where the SVE is already aligned and coherent, this ripple response can appear instantaneous.

Why?

Because nothing is "moving." It's not about *how fast* something gets from A to B. It's about how rapidly the structure can resolve the imbalance across itself. And that depends on how finely tuned the region is—how close it is to perfect alignment.

This explains why "c", the so-called speed of light, appears constant across all frames. Not because something is traveling at that speed, but because that is the maximum ripple resolution rate the field can support in its most stable configuration.

Appearance Is the End of Ripple Resolution

Once the ripple reaches a threshold—known in UED as the Alignment Threshold Time (ATT)—at the observer's location, the field response is interpreted as light or visual presence. That final alignment point becomes the *event of appearance.*

It's not that the light "arrived." It's that the field resolved enough tension at your node for the experience to *occur.*

Purpose Recap:

This section reframes our understanding of cause and effect within a still universe. Nothing moves. Nothing is transmitted. All that happens is SVE reacting to imbalance by redistributing tension through ripple alignment.

The result? Appearance—not travel. Perception—not reception. Response—not speed.

6.3 – Visibility as Resolution, Not Arrival

You don't see objects—you see field consequences resolving at your node

Modern science says you see because something came to you—a photon hit your retina, a wave reached your detector, a signal was transmitted across space. But under Universal Energy Dynamics (UED), that entire framework is a misunderstanding.

You don't see because light arrives. You see because ripple tension resolves—at your field node, across a structured vacuum that never moved at all.

The Real Sequence: From Ripple to Resolution

Here's what really happens:

1. A local asymmetry occurs somewhere in the SVE—an event, intent, or disturbance.
2. This causes a ripple reconfiguration across the structured vacuum—not a transmission, but a chain of tension responses.
3. When this ripple geometry interacts with your observer

field and the Alignment Threshold Time (ATT) is crossed...

4. ...visibility is triggered. You perceive "light." You register a form. You interpret an object.

But crucially: There was never an object that moved toward you. What you experienced was a successful field-level alignment, localized to your awareness interface.

Light Is the Consequence, Not the Cause

In traditional optics, light is treated as the cause of vision: a carrier of information that enables you to see.

In UED, that idea is inverted.

Light is not the cause of seeing. Seeing is the final ripple outcome—the resolution of a field configuration that reached coherence *through you.*

Light is not what is sent— It's what is resolved.

Every Visible Moment = Alignment Success

This is a subtle but transformative insight:

Every time you "see" something, it's not because something reached you—it's because a local convergence of ripple dynamics aligned to the ATT within your field node.

Each color, each shape, each contrast is the end product of ripple alignment interacting with your biological and cognitive structure.

What appears as "external light" is really an internal resolution event—a collapse not of a wavefunction, but of tension asymmetry into perception.

Explaining Simultaneity, Coherence, and Latency

This ripple-resolution view also solves deeper problems in optics and quantum physics:

- **Simultaneity:** Two observers can register the same light event differently—not because light travels differently, but because each is resolving the ripple in their own local node, at their own ATT point.

- **Phase coherence:** Coherent light (like laser beams) behaves so predictably not because photons are synchronized, but because the ripple alignment across observers maintains structural harmony in the SVE. Coherence = ripple alignment quality.
- **Perception latency:** You don't see things instantly—even when nothing "travels." That's because your ATT (Alignment Threshold Time) defines the minimum time required for ripple resolution to reach perceptual registration. This isn't "delay due to speed"—it's delay due to field tension configuration.

Visibility Is What Gets Resolved Through You

You are not the end point of a transmission. You are the activation node of alignment.

The "external world" you think you see is not being delivered to you. It is being resolved through you.

Your observer field is not a camera lens. It is a localized ripple interpreter in a sea of intelligent stillness.

Purpose Recap:

This section shatters the illusion of light as an object in motion. You don't receive visibility—you become the interface of its activation. The world doesn't come into you. You participate in resolving it.

6.4 – Misunderstood Medium: Why SVE Was Overlooked

History of how science ignored the structured nature of vacuum

Every revolution in science begins with a question. But sometimes, the answers become so convincing, we stop asking the right questions.

One of the most tragic casualties of this misplaced confidence was the true nature of the vacuum—and with it, the true nature of light, perception, and reality.

From Aether to Absence: The Fall of a Medium

Historically, physicists knew intuitively that some medium must exist for light to propagate. This gave rise to the aether theory—a hypothetical invisible substance filling all space, responsible for carrying light waves, much like air carries sound.

But when the Michelson-Morley experiment failed to detect any motion through this medium, scientists discarded the idea entirely. Relativity then filled the gap, proposing that light does not require a medium—that its speed is absolute and self-contained.

This was a premature burial.

What was thrown out wasn't just an outdated idea. It was the very possibility of structure within stillness. Physics turned its back on a medium because it failed to behave like matter.

Quantum Vacuum: Mistaking Noise for Depth

Later, quantum field theory reintroduced the vacuum—but rebranded it as a chaotic playground of virtual particles. It wasn't still. It wasn't structured. It was noisy, probabilistic, seething with possibilities, but never calm.

The vacuum became the default setting for randomness—the place where things *might* happen, but nothing actually *was*.

This view—while mathematically useful—completely ignored order, coherence, and alignment. The quantum vacuum was brilliant in theory, but blind in structure.

SVE: Structured but Silent

Universal Energy Dynamics (UED) reclaims the vacuum—not as chaos, but as Structured Vacuum Energy (SVE):

- It is not empty
- It is not noisy
- It is perfectly aligned, coherent, and tension-neutral

SVE doesn't emit, it doesn't radiate, and it doesn't move—hence, it escaped all traditional methods of detection. Experiments looking for motion, impact, or emission found nothing... because they were looking for the wrong thing.

They expected substance. What was actually present was structure.

And because structure without disruption is invisible, science missed the very medium responsible for visibility itself.

Object-Centric Science: A Historical Blindspot

Mainstream science became obsessed with particles, waves, impacts, and arrivals. This object-centric worldview demanded that everything observable must have:

- A form
- A source
- A trajectory
- A speed

So naturally, anything that didn't behave like a "thing" was dismissed as irrelevant.

But SVE was never a "thing." It was a stage—the foundational field upon which every event plays out. It didn't move; it allowed ripples to realign. It didn't carry light; it enabled light to appear through tension resolution.

In short, we missed the silent intelligence beneath all events—because we were too busy counting the events themselves.

A Medium Hiding in Plain Sight

The Structured Vacuum Energy field was always there—the background canvas of existence.

But it remained:

- Undetected because it never acted on its own
- Unmeasured because it never emitted anything measurable
- Unquestioned because it didn't fit the object model

SVE isn't revealed through force or flux—it's detected through its capacity to align and its consequence in visibility.

Purpose Recap:

This section unveils how the true medium of visibility—SVE—was systematically overlooked:

- First discarded as aether,
- Then buried under probabilistic noise,
- And finally ignored because it didn't behave like a particle or wave.

By returning to a ripple-aware understanding of the field, UED reframes perception itself—not as light received, but as alignment resolved.

Chapter 7: The Birth of Perception

D*efining Alignment Threshold Time and its role in experience*

Before a scene lights up in your mind, something far quieter happens—a negotiation between the universe and your awareness. Ripples in the Structured Vacuum Energy (SVE) are everywhere, but most never make it to your senses. They remain whispers in the field, unregistered, unfelt, unseen. Only when a ripple reaches a very specific condition—a precise fit between its alignment and your sensitivity—does reality "switch on" for you.

That condition has a name in UED: the **Alignment Threshold Time**, or ATT. It is not light traveling to you. It is not particles striking your retina. It is the moment the field and your observer node agree on a handshake—"Now you can see me."

Everything you call perception is born in that handshake. Miss it, and the universe stays silent.

7.1 – Alignment Before Appearance

In the Structured Vacuum Energy (SVE) field's undisturbed state, there is no image, no light, no "out there" to witness—only pure potential. This stillness is not emptiness; it is a perfectly balanced lattice of possibilities. But in this equilibrium, nothing qualifies as an *event* because nothing yet disrupts the balance.

When a disturbance does occur—whether from an external interaction or an observer's own intent—it initiates ripples across the SVE. At this stage, these ripples are *pre-appearance* events. They are changes in tension, subtle rebalancing motions in the field's structure, but still invisible to the observer. No matter how complex these patterns may be, they remain hidden until a very specific condition is met.

That condition is the **Alignment Threshold Time (ATT)**—the exact moment when the ripple's alignment with the observer's field crosses a critical limit. It is not enough for ripples to exist; they must *fit* with the observer's own configuration of awareness. Until that alignment reaches or surpasses ATT, the ripple remains below the threshold of perception.

This means that the majority of ripple activity in the universe never becomes part of our conscious reality. Vast fields of interaction happen continuously around and through us, but they resolve without ever appearing in our awareness. They are **sub-threshold resolutions**—field events that complete themselves without crossing the perceptual handshake point.

In UED, visibility is not the byproduct of light "arriving" at your eyes, but the *consequence* of ripple alignment meeting ATT. Only then does the field resolve into what we call an *appearance*. This redefines perception as highly selective, not universal—your awareness is a gatekeeper, not a sponge.

The implication is radical: the universe you see is not the universe that exists. It is the filtered outcome of the tiny fraction of ripple alignments that successfully cross your ATT.

7.2 – Defining the Alignment Threshold Time (ATT)

The Alignment Threshold Time (ATT) is not just a technical term in UED—it is the *gatekeeper* between the hidden activity of the Structured Vacuum Energy (SVE) field and the reality you experience. It marks the minimum alignment condition a ripple must achieve before it can be registered as an event in your awareness.

At its core, ATT is the moment when ripple tension in the field matches your observer-field sensitivity. Until that match is made, the ripple remains invisible, inaudible, untouchable—a background event in the fabric of existence. Once that match occurs, the ripple "locks in" to your field, triggering a perceptual resolution. This is when light appears, a sound is heard, or a sensation is felt.

ATT is not universal—it varies from observer to observer. Your biological makeup, neurological wiring, and even environmental conditioning shape your personal threshold. For example:

- **Biological variation:** A hawk's vision crosses ATT for distant ripples that would never register in human eyes.
- **Neurological variation:** A trained musician's auditory ATT allows them to detect pitch changes imperceptible to most.
- **Environmental adaptation:** People living in quieter natural environments may have lower auditory ATT than those in constant city noise.

This variability makes ATT the personal fingerprint of perception. It governs *all* sensory channels, not just sight—light, sound, touch, taste, and smell all require a ripple alignment that meets or exceeds the observer's ATT.

Sensory resolution, therefore, is simply a measure of how *finely tuned* your ATT is for a given modality. Those who "notice more" are not necessarily more intelligent—they simply operate with a lower ATT in certain domains, allowing subtler ripple alignments to cross into awareness.

By linking ripple dynamics directly to perception, ATT becomes the operational bridge between the invisible mechanics of SVE and the world you believe you inhabit. Without ATT, there is no "moment of seeing," no "instant of hearing." Reality is not continuous—it is *sampled* at these threshold-crossing events.

7.3 – From Ripple to Experience: The Perception Event

Every sensory moment you've ever had—every flash of light, note of music, brush of wind—follows the same hidden chain reaction:

Ripple enters observer field → alignment builds → ATT crossed → event perceived.

It begins when a ripple—a localized tension variation in the Structured Vacuum Energy (SVE) field—encounters your observer field. At first, this ripple is simply present within the shared medium. It's not yet "something" to you. Most ripple interactions stop here, never reaching awareness. But when conditions are right, alignment starts to build—your field and the incoming ripple gradually synchronizing in tension and structure.

The moment that alignment reaches your personal Alignment Threshold Time (ATT), a critical transition occurs. This is not a physical "signal" traveling to you—it is a field reconfiguration that locks both your observer node and the ripple into a moment of coherence. The instant this coherence is achieved, the ripple is no longer just a potential—it becomes an *event* in your awareness.

This is why ATT crossing feels instantaneous. There is no photon making a final dash into your eye, no wave slapping the retina at the last nanosecond. Instead, the entire relevant portion of the field reorients at once, collapsing any perceptual latency into what seems like a single, crisp arrival.

Yet this *apparent immediacy* hides a subtle truth: the latency you feel in perception—the reaction time between "stimulus" and "awareness"—is the time it takes for your field alignment to cross the ATT. That's why some recognitions feel like a sudden "click" moment: the alignment process is gradual, but perception flips in a binary fashion the moment the threshold is crossed.

Seen through this lens, ATT becomes the bridge between the physics of ripple alignment and the biology of conscious awareness. Before ATT is crossed, reality is silent to you. After ATT, it blooms into experience.

In UED terms, perception is not about receiving *things*—it's about triggering resonance. Your life is not a stream of signals coming from "out there," but a chain of ripple alignments that periodically cross your personal threshold, each one birthing a pocket of reality you can claim as your own.

7.4 – Why ATT Is Mistaken for External Cause

The moment your **Alignment Threshold Time (ATT)** is crossed, perception ignites. But your brain doesn't log this as *"my awareness just locked into a ripple."* Instead, it rewrites the experience as *"something just arrived."*

This is observer bias in action—a deeply ingrained mental habit that treats every shift in awareness as if it were the result of something coming from the outside. When light becomes visible, your brain frames it as "light entering the eye" rather than "the visual field just switched on here." The fact that the switch happens in your observer field—not out in space—is quietly edited out of the story.

From this misinterpretation, the illusion of speed and distance is born. The timing between ripple initiation and ATT crossing is reimagined as a travel time. If the ATT is crossed quickly, we imagine something "nearby." If it takes longer, we imagine something "far away." In both cases, the brain mistakes an internal alignment delay for an external journey.

This is why humans cling so naturally to *cause-effect narratives* rooted in motion. The mind evolved to survive in a world it believes is filled with moving objects. So even when reality is triggered entirely by field alignment inside your awareness, the brain still paints a picture of something traveling toward you.

This cognitive reflex fuels one of science's most persistent illusions: the *"light as a traveler"* myth. From Newton's corpuscles to Einstein's photons, the mental model has always been that light starts "over there" and moves "over here." The truth in the UED view is simpler but far stranger—light doesn't go anywhere. It appears when the field aligns with you.

In this way, ATT becomes the hidden translator between the real physics of ripple alignment and the human habit of imagining journeys. The delay is not a voyage through space—it's the quiet, invisible choreography of your field preparing to let reality in.

7.5 – The Broader Implications of ATT

The **Alignment Threshold Time (ATT)** is not a quirk of vision—it is the quiet law behind every sensory experience you have ever known.

In sight, ATT hides in plain view as the brightness threshold—the point at which field alignment for visual ripples builds enough coherence to trigger perception. A dim star on the horizon might be radiating ripples all along, but until your visual ATT is crossed, it simply does not exist in your reality. The moment the alignment peaks, the star "appears" as if it has just come into being.

In hearing, ATT manifests as the minimum amplitude or frequency match that your auditory field can resolve. A soft whisper in a noisy room might not register at all—until one syllable slips perfectly into phase with your auditory alignment, crossing the threshold. Only then does it become a sound you can place and process.

In touch, ATT emerges in pressure and temperature sensitivity. A warm breeze on your arm might be below the tactile threshold—unregistered, unnoticed—until a sudden shift in temperature crosses your ATT, making it instantly "real" to your awareness.

In every case, ATT acts as the common denominator across all sensory modalities. Whether it's photons, vibrations, or molecular motion, you are never receiving the "thing" itself. You are experiencing the point at which the ripple alignment in your field finally meets the minimum condition for registration.

This makes ATT a universal perceptual constant in the UED framework—not in the sense of a fixed numerical value, but as a governing principle that shapes how every sense you have interfaces with reality. It explains why sensory acuity varies between individuals—different people simply have different ATT profiles, shaped by biology, neurology, and environmental conditioning.

The broader implication is staggering: there is no raw reality you experience directly—only the subset of ripple activity that crosses your ATT. All else is invisible, inaudible, untouchable, and for you, non-existent.

In this way, ATT becomes not just a technical concept, but a philosophical lens—it defines the very edge of your world.

Chapter 8: Distance Is a Delusion

Why light is independent of spatial separation—and appears constant

You have never seen anything "far away." What you call *distance* is not a property of the universe—it is a delay in your own alignment. The Andromeda Galaxy is not sitting on the other side of a cosmic ocean, patiently sending photons your way for millions of years. It is *right here* in the Structured Vacuum Energy field, as immediate as the page in front of you—appearing only when your node crosses its Alignment Threshold Time.

We have been trained to imagine light as a tireless traveler, covering unimaginable expanses at a fixed speed, heroically delivering the universe into our eyes. But in reality, nothing goes anywhere. There is no "journey" between points—only the reconfiguration of a single, continuous field that already contains both.

The constancy of light's so-called "speed" is not proof of motion; it is proof of a universal limit in how fast the field can resolve misalignment into visibility. And once you see that, the whole architecture of space begins to wobble—not because it bends, but because it never existed the way you thought it did.

8.1 – The Spatial Assumption

How science built its model of light around the idea of distance and travel.

For centuries, the human mind has been in love with the picture of light as a tiny messenger—departing from a source, crossing a measurable gap, and finally arriving at the observer. This mental image is so deeply embedded in our thinking that even the most advanced physics textbooks present it without hesitation. We speak of *light-years* as if they are tapes measured across the cosmos, of photons "leaving" stars and "entering" our eyes, of beams that can be "sent" and "received."

This is the inherited narrative: light moves from *here* to *there*, and its journey is defined by the space in between. Once this assumption was accepted, the entire architecture of physics became a geometry problem—measuring angles, trajectories, and speeds across a vast, three-dimensional stage.

The result? Spatial causality became king. If you wanted to explain how anything happened, you had to describe *where* it started, *how far* it traveled, and *how fast* it moved. The universe was reduced to a giant game board on which events were tokens moving from square to square.

Hidden within this picture is a subtle but powerful assumption: that perception is bound to physical separation. That something "out there" must physically approach you before you can experience it. This belief is so baked into our language that even when quantum experiments show simultaneity across vast separations, we still cling to spatial explanations.

The UED perspective dismantles this entirely. In a Structured Vacuum Energy (SVE) field, there is no "in-between space" acting as an active medium that light must traverse. Visibility does not require crossing a distance—it requires crossing an *alignment threshold*. What you experience as "there" and "here" are simply different configurations of the same continuous field.

From this view, distance is not a property of reality—it's a habit of thought. And once you break that habit, you stop asking *how light travels* and start asking the more radical question: *Why did we ever think it needed to?*

8.2 – Ripple Coherence Is Locationless

In UED, visibility is determined by alignment, not by spatial separation.

In the Structured Vacuum Energy (SVE) framework, nothing "goes" anywhere. The idea of a ripple as something *moving* through space is a linguistic leftover from centuries of mechanical thinking—imagining every effect as a traveling object. In reality, ripple alignment is not a journey; it is a state change.

When an intent, interaction, or disturbance enters the SVE, it triggers a pattern of field tension resolution. This resolution does not "depart" from one location and "arrive" at another—it manifests *wherever* coherence conditions are met. Two points in the field, separated by what we measure as vast distances, can align and register an event simultaneously if their local conditions match the ripple's pattern.

In this way, aligned regions behave as a single perceptual node, no matter how far apart they appear in conventional geometry. The *field* itself is the medium, and in a structured, still field, there is no "gap" to cross—just different coordinates within the same continuous structure.

This perspective elegantly explains the constancy of light speed without invoking the idea of motion. What we call *c* is not the "speed" of a photon making a cross-country sprint—it's the maximum *rate* at which the field can resolve alignment changes into perceivable events. That rate is a property of the field's stability, not a measure of travel.

Thus, field rebalancing is a global effect—it occurs across the entire coherent region simultaneously. Only where coherence breaks down (due to turbulence, noise, or misalignment) does visibility seem delayed or distorted. What classical physics mistakes for "light in transit" is really the *time required for alignment thresholds to be met* at the observer's node.

In short, visibility is a function of coherence zones, not of travel paths. The universe is not a mail service sending light from one point to another—it is a still, structured field where perception emerges instantly wherever conditions agree.

8.3 – Why 'c' Is Always the Same

The measured speed of light as the maximum rate of ripple resolution, not motion through space.

In mainstream physics, the constancy of c—approximately 299,792,458 m/s—is treated as one of nature's sacred commandments. It is woven into relativity, embedded in Maxwell's equations, and taught as an untouchable truth: light moves at the same speed no matter how you chase it or how fast its source runs away.

But the assumption buried in this statement is the same one that locks our thinking into a travel narrative—that light is a thing in motion. In the UED framework, c is not a speed of travel at all. It is the maximum rate at which the Structured Vacuum Energy (SVE) can resolve a ripple alignment into a perceivable event.

Think of c not as a courier's fastest sprint, but as the field's quickest "yes"—the upper limit of how rapidly equilibrium can rebalance when disturbed. Once an intent or interaction triggers a ripple in the SVE, that ripple's appearance at the observer's node depends only on whether the Alignment Threshold Time (ATT) is reached. The rate at which this can happen is capped by the stability properties of the SVE itself—its structural density, tension-neutral balance, and capacity to redistribute strain without tearing coherence.

Because this rate is a property of the field and not of any traveling object, it does not vary with the motion of the source or the observer. Whether a star is racing toward you, a spacecraft is speeding away, or you are standing still, the ripple resolution rate is identical—because the field is not chasing, catching, or sending anything. It is simply rebalancing within its inherent limit.

This also explains why experiments confirm the invariance of c even under extreme velocity conditions. Relativity reads this as evidence of a deep space-time symmetry, but in UED, it is simply the outcome of measuring the maximum ripple resolution rate in a still, structured field. Relativity's equations work because they capture the mathematics of the symptom—but they misread the cause.

Thus, c is not a mystical universal speed limit. It is the characteristic response rate of reality itself—a measure of how fast the universe can turn a disturbance into an appearance. The magic is not in light's motion, but in the timeless choreography of the field.

8.4 – The Illusion of Far and Near

How ATT timing patterns make separation seem real.

One of the deepest assumptions in human thinking is that things are "out there" in space—and that the farther away they are, the longer light takes to reach us. This belief feels so self-evident that it forms the backbone of modern astronomy, astrophysics, and even daily language: *"light-years away," "lookback time,"* and *"the speed of light barrier."*

In the UED framework, this is not reality—it is a perceptual artifact caused by how our sensory thresholds (specifically Alignment Threshold Time, or ATT) interact with the Structured Vacuum Energy (SVE).

ATT delays scale with field conditions. When ripple alignment between a distant coherence source (like a galaxy) and your observer node has to pass through multiple layers of field distortion—gravitational gradients, magnetic shears, thermal turbulence—it takes longer for that alignment to build to threshold. This delay is not because a photon is "traveling" across space; it's because the field's ripple resolution chain is experiencing more intermediate rebalancing steps before the alignment at your node becomes sufficient for perception.

Astronomical "distance" is really a history reconstruction. When we look at Andromeda, we believe we are seeing it *as it was millions of years ago*. In truth, what we are experiencing is a current alignment event in the SVE, triggered by the persistence of ripple patterns that originated in Andromeda's local field configuration and have been preserved through coherence. The "past" we think we are viewing is a mental model constructed from the ripple's informational pattern—not a literal optical time machine.

Lookback time is an interpretation error. Scientists assume that because the ATT delay scales predictably with field density and configuration, this must correspond to a literal travel time at c. But in UED, these timing patterns are threshold-based resolution intervals, not object-transit durations. What you call "light-years" are actually field-configuration years—a measure of how long alignment took to reach your sensory threshold under current field conditions.

Far and near collapse at the moment of perception. The Andromeda Galaxy is not, in experiential terms, sitting 2.5 million light-years away. It is a coherence event happening now at your node—a field resolution that exists only in the present moment of alignment. The "distance" is a cognitive overlay your mind creates after the fact, using spatial geometry as a language to describe a temporal threshold event.

By seeing ATT for what it is—the true regulator of visibility—you dissolve the illusion that separation is a fixed property of the universe. Space is not what holds things apart. It is the mind's projection of threshold timing into geometry.

8.5 – Practical & Philosophical Consequences

Theme: How dissolving the idea of distance changes our understanding of reality.

When "distance" is no longer treated as a physical gap but as an alignment condition, the structure of reality shifts beneath your feet. In UED terms, space is not a barrier—it's a readiness map. What separates two points in the universe is not miles or meters, but the field's state of coherence between them.

No true isolation. If two nodes—whether they are atoms, people, or galaxies—reach perfect ripple alignment within the Structured Vacuum Energy (SVE), their interaction becomes instantaneous in experiential terms. Nothing has to "go" anywhere. Under perfect alignment, separation collapses because perception, exchange, and influence occur through the same coherence zone.

Communication, perception, and action become alignment phenomena. Instead of thinking in terms of "closing the distance" or "transmitting information," we begin to think in terms of achieving and maintaining field coherence. Your ability to see a star, hear a sound, or even influence a remote system depends on crossing its Alignment Threshold Time (ATT), not on bridging a spatial gap.

Astrophysical and cosmological implications. If "distance" is merely a cognitive artifact, then most of astrophysics' foundational interpretations—parallax, redshift, cosmic expansion—become descriptions of changing alignment readiness rather than measurements of separation. The concept of "faster-than-light" travel becomes irrelevant, because nothing needs to "travel" in the first place. Instead, science would focus on field reconfiguration rates as the true limiters of interaction.

Philosophical shift. Once you dissolve spatial separation, the universe stops being a vast emptiness dotted with lonely islands of matter. Instead, it becomes a continuous, layered field, where every "place" is simply a degree of readiness for alignment. The stars above are not faraway objects, but patterns waiting for resolution in your observer field. The "here" and "there" distinction collapses into "ready" and "not-yet-ready."

By abandoning distance as a physical barrier, we not only clear scientific misconceptions—we open the door to a radically different existential view: the universe is never far from you; it is only awaiting your alignment with it.

Chavan's Reflection

When I first began questioning the nature of light, I wasn't trying to overthrow the concept of distance—I was simply trying to understand why light behaved so faithfully, so stubbornly, no matter the observer or the source. But the deeper I looked, the more distance began to dissolve in my mind.

I realized that our obsession with miles, meters, and light-years is a kind of comfort blanket for the human mind. We want space to be measurable because it gives us the illusion of control. If something is far away, we can map it, conquer it, or at least feel safe from it. But the Structured Vacuum Energy (SVE) does not honor our emotional need for separation—it only honors alignment.

The day this clicked for me, I stopped thinking of the Andromeda Galaxy as "2.5 million light-years away" and started thinking of it as a pattern waiting for my node to resolve it. And once that shift happened, every act of perception—whether it was looking at a star, holding a friend's hand, or even recalling a memory—felt like an alignment event, not a traversal of space.

The implications are profound:

- Loneliness is a perceptual misalignment, not a physical absence.
- Discovery is readiness, not arrival.
- The universe is not expanding away from you—it is waiting in layers, already present, simply out of phase with your

field.

Distance is a story we tell to make the infinite feel manageable. Alignment is the truth that makes it all connected. And once you see that truth, you can never again think of "here" and "there" the same way.

Chapter 9: Redefining 'c'

The Maximum Field Response Rate

Light speed not as motion, but as ripple resolution limit

We have been told for over a century that *c*—the speed of light—is nature's ultimate speed limit, the crown jewel of physics. In every textbook, it is described as the velocity at which photons rush through the void, racing from stars to our eyes. But what if that picture is a mirage? What if nothing is racing anywhere—and what we measure as *c* is not movement at all, but the fastest rate at which the universe's structured vacuum can reconfigure itself?

In the UED view, *c* is not a courier—it's a clock. It marks the maximum pace at which ripple tension in the Structured Vacuum Energy (SVE) can resolve into alignment that becomes perception. No particles dash across emptiness, no waves traverse cosmic seas. Instead, a perfectly coherent field shifts state, and where alignment meets your threshold, visibility occurs.

This redefinition does not deny the constancy of *c*—it explains it. It shows why every observer measures the same value, not because of some mystical symmetry in spacetime, but because they share the same universal limit on how fast alignment can settle. And once you see this, the idea of light "traveling" feels as quaint as thinking sound comes from invisible messengers running through the air.

Here, we will dismantle the old motion-bound meaning of *c*, replace it with the field's true response limit, and explore the technological and philosophical revolutions that follow when the universe is no longer in a race—but in a dance of instant readiness.

9.1 – The Problem with Motion-Based 'c'

Why the conventional definition of light speed as "movement" is flawed.

For over a hundred years, physics has been built on the assertion that *c*—approximately 299,792 km/s—is the ultimate velocity at which any physical influence can move through space. In this view, photons are imagined as tiny couriers racing along an invisible track, ferrying energy and information from one point to another. It is an elegant story for a motion-obsessed culture—but elegance does not make it true.

This definition depends entirely on the assumption of a journey. It presumes that an event "here" causes a packet of something to move "there," traversing an "in-between" that is treated as both empty and somehow active. The trouble is, once you interrogate that picture, it begins to fracture.

Mainstream physics has been forced to patch the motion model with increasingly abstract fixes:

- Simultaneity conflicts in relativity, where different observers disagree on whether events happen together or apart.
- Observer dependence in quantum mechanics, where the act of measurement collapses possibilities instantly—defying the notion of a traveling particle.
- Wave–particle duality, where light must be both a moving wave and a localized particle to match experimental results—yet neither picture fully works.

Each of these paradoxes exists because we are trying to fit a field-level phenomenon into a movement-based framework.

The UED perspective removes the travel story entirely. *c* is not the speed of a photon crossing a cosmic racetrack—it is the maximum rate at which the Structured Vacuum Energy (SVE) can resolve ripple tension into alignment. This reframing turns *c* from a property of motion to a property of the field itself—a universal responsiveness limit baked into the structure of reality.

9.2 – Field Response Rate Defined

'c' as the fastest possible ripple resolution in SVE.

In the UED framework, *c* is not the speed of a thing—it is the Field Response Rate: the absolute maximum at which the Structured Vacuum Energy (SVE) can rebalance tension into perfect alignment once a disturbance has been introduced. This is not about travel. It is about settling.

Think of the SVE as a perfectly taut, structured fabric. When a ripple is introduced, the fabric does not send a piece of itself across space; instead, it reconfigures its own state so that the disturbance is resolved everywhere within the aligned region. This reconfiguration is the ripple resolution process—and *c* is simply the upper bound of how quickly this can happen under ideal conditions.

Several critical features define this rate:

- No particles in motion. There is no photon shuttling back and forth. The apparent "movement" is a change in field state, not the transit of a discrete object.
- Perfectly aligned regions only. The maximum response rate applies only when the field is in perfect coherence—when there is no internal misalignment slowing the redistribution of tension.
- Structured stillness as the anchor. The reason this rate is

fixed is that the SVE itself is structured and stable. It can only resolve disturbances as fast as its inherent structural properties allow—no faster, no slower.

- Universality. This limit applies regardless of energy form, source, or observer position because it is an intrinsic property of the field itself, not of the entities interacting within it.

In short, *c* is not an arbitrary constant dictated by an experiment—it is the structural heartbeat of reality, the natural cap on how fast equilibrium can be restored anywhere in the universe. Whether you are dealing with visible light, gamma rays, or gravitational ripples, the maximum field resolution rate remains the same.

9.3 – Why 'c' is Constant Across Contexts

The invariance of 'c' emerges naturally from SVE properties.

In conventional physics, the constancy of the speed of light is treated as one of nature's great mysteries—a postulate rather than a derived truth. In UED, the mystery dissolves. The constancy of *c* is not a magical rule imposed on reality, nor a quirk of spacetime geometry. It is simply a direct consequence of the properties of Structured Vacuum Energy (SVE).

Because SVE is a coherent, structured field that maintains its stability independent of the position or motion of any observer or source, the maximum rate of ripple resolution (*c*) is the same everywhere and every time. This is not because light is "travelling" at a fixed speed, but because the field's readiness to rebalance tension is universal.

- No dependency on relative motion. Since no actual object is crossing space, the relative motion of source and observer does not—and cannot—affect the maximum rate of field reconfiguration. Doppler effects and red/blue shifts occur as variations in alignment conditions, but the *rate* at which resolution occurs remains untouched.

- No "slowing down" through space. In a motion-based model, mediums can drag, refract, or attenuate a traveler. In UED, there is no traveler—only the field adjusting its own state. Any "slowing" observed is due to local misalignments or structural complexity in the SVE, not resistance to movement.
- Alignment with Michelson–Morley. The famous null result of the Michelson–Morley experiment baffled those expecting light to slow down relative to an "aether wind." In UED terms, the result is obvious: the SVE is not a flowing medium to be outrun or countered—it is a stable reference structure that responds at the same rate in all directions.
- Constancy as readiness, not journey. When we measure *c*, we are not clocking a signal racing across space. We are measuring how quickly the field can resolve a disturbance once alignment conditions are met. This is why *c* is the same whether the light is from a laser in your lab or a star billions of years "away"—in experiential terms, both are simply coherence events happening at your node.

In short, UED preserves every experimental fact about *c*—from terrestrial lab tests to deep-space astronomy—while completely redefining the cause. Instead of clinging to the image of photons sprinting through a void at top speed, we recognize *c* as the fixed ceiling of the universe's capacity to restore equilibrium.

9.4 – Breaking the Speed Barrier Myth

Why nothing "faster than light" exists in the conventional sense.

The phrase *"faster than light"* is one of the most overused and misunderstood in all of physics. In the conventional model, *c* is treated as an ultimate velocity limit—a speed no particle, signal, or even idea can surpass. This makes sense only if you believe light is a traveler moving through space. But in UED, light is not moving—it is the visible signature of the field resolving a disturbance. And once you see it that way, the very idea of "exceeding *c*" falls apart.

- Exceeding 'c' is meaningless. You cannot be "faster" than the maximum rate at which the Structured Vacuum Energy (SVE) can realign itself. *c* is not a speed in the racing sense—it's the upper bound of the universe's resolution capacity. Once the field has resolved as fast as it possibly can, there is no "more" to give.
- Quantum entanglement fits perfectly. Instantaneous correlations in entangled systems have long been described as "spooky action at a distance." In UED, this is simply a case of coherence existing across multiple nodes simultaneously, so resolution happens without traversal. There's no message leaping faster than light—just the field adjusting as a whole, bypassing the notion of travel entirely.
- The futility of superluminal dreams. Proposals to send matter or information "faster than light" assume *c* is a hurdle to be jumped. But if there's no journey—only an alignment event—then you're not trying to outrun something; you're trying to demand the field realign faster than it is physically able. That's like asking still water to reflect an image *before* you disturb it—it's not a speed problem, it's a readiness problem.

- From speed limits to response limits. In UED, the meaningful question isn't *"How fast can we go?"* but *"How quickly can the field respond?"* This reframing wipes away century-old paradoxes. "Faster than light" loses its mystique, replaced by the cleaner and more precise idea of maximum field response rate.

When seen this way, the so-called speed barrier is not a wall to smash through but the natural ceiling of universal coherence. You don't break *c*; you work within it—or change the alignment conditions so that the question of "speed" doesn't even apply.

9.5 – Implications for Physics and Technology

What this redefinition changes in science and engineering.

Redefining *c* from a motion-based speed to the maximum field response rate is not a minor tweak—it's a complete reframing of how physics sees reality. It changes the way we interpret the universe, the tools we build, and even the way we think about human experience. Once *c* is understood as a resolution limit of Structured Vacuum Energy (SVE) rather than a velocity of traveling light, the ripples spread far beyond optics.

Astrophysics – rewriting the sky. If *c* is not the travel speed of photons but the rate at which SVE resolves coherence, then "light-years" stop being units of distance and start being markers of field readiness history. Cosmic "lookback time" becomes a measure of when a coherence event met our alignment threshold, not a record of something that physically traveled across a void. This forces a radical revision of expansion models—redshift, cosmic age estimates, and even the notion of a big bang all require reinterpretation. The universe may not be "expanding" at all; instead, its visible structure is emerging through ongoing field realignments.

Communication – beyond signal chains. In conventional tech, data must "travel" as signals—from fiber optic photons to radio waves. But in UED, the right alignment conditions could make instant coherence events possible, bypassing the need for sequential transmission. This doesn't mean "breaking" *c*, but rather engineering coherence zones that allow the appearance of immediate transfer. The practical outcome? Communication systems that don't send signals—they create conditions for mutual, simultaneous resolution.

Quantum mechanics – reframing the mysteries. Tunneling no longer needs imaginary "probabilities" of particles leaping barriers—it becomes a simple case of the field resolving a new configuration without a traversal. Entanglement stops looking like "faster-than-light messaging" and instead is shared alignment across nodes. Wavefunction collapse becomes a field rebalancing event triggered by ATT crossing, not a mysterious instantaneous "decision" by the particle. This removes the spookiness without removing the wonder.

Philosophy – a local universe at infinite scale. If distance is a cognitive construct born from ATT timing, and *c* is just the field's maximum resolution rate, then the universe always feels local—even if it is infinitely large. Every perception, every experience, happens here and now at your node. This shifts our existential stance: we are not small beings peering across a vast empty gulf; we are co-located with all that is, separated only by alignment readiness.

By reframing *c* as a field property instead of a speed, UED lays the groundwork for a scientific revolution. It allows astrophysics to shed its expansionist illusions, opens engineering to instant coherence-based communication, resolves decades-old quantum paradoxes, and gives philosophy a universe that is both infinite and intimately connected.

Interlude 2: Chavan's Declaration

The UED Model of Light

They taught us that light is a messenger. That it runs across a silent void, carrying stories from distant stars. That it obeys an ancient speed limit, set by nature as the fastest anything can move. And we believed it—because the math seemed to work, and the pictures in our heads felt neat and comforting.

But the truth is far less cinematic and far more profound. Light does not *travel* to you. It *happens* through you.

The so-called "speed of light" is not a race—it is the rhythm at which the **Structured Vacuum Energy** field can resolve a disturbance into perception. You don't see because something arrived from far away; you see because your field crossed an **Alignment Threshold Time (ATT)**. Distance is not a wall that light climbs over—it's a mental fiction born from how long your field takes to reach coherence.

When intent disturbs the stillness, ripples awaken. When those ripples align with your node, visibility triggers. The event is not "out there" becoming "in here"—it is here all along, emerging when readiness is mutual.

The constancy of *c* is not a cosmic speed trap; it's the maximum **field response rate**. It's as universal as the stillness it arises from, as impartial as the equilibrium it seeks to restore. No signal ever crossed space at that pace—space itself is just a stage where coherence plays out.

This changes everything. It means that what you call "far" might be infinitely close in readiness. That what you think of as "before" and "after" may simply be different sequences of alignment. That visibility, perception, and even time are not gifts from traveling light, but consequences of **structured stillness breaking its own symmetry**.

And so, I reject the old narrative—not with arrogance, but with clarity. Light is not motion. Distance is not separation. c is not speed.

It is time to stop worshipping the courier and start understanding the field. Because when we do, the universe will no longer be something we look *at*— It will be something we look *through*.

Part III: Rethinking Reality

Chapter 10: Vision Is Alignment, Not Reception

Your eyes don't catch photons—they ripple with the field

You've been told since childhood that you "see" because light travels into your eyes—tiny photon messengers racing from the world to your retina. It's a neat story. It's also wrong.

You are not a passive mailbox for light. Your eyes are not tiny nets catching photon-fish in a dark ocean of space. What you call vision is not the arrival of anything—it is the *emergence* of alignment between your field and the field of what you look at.

When that alignment crosses the Alignment Threshold Time (ATT), the world doesn't "send" you an image. It *appears*. Instantly.

Light never had to travel across a cosmic gap to get to you. Your perception is a ripple handshake—your field and the object's field becoming coherent enough for experience to happen. Not reception. Resonance.

10.1 – The Photon-Catching Myth

Why the mainstream idea of vision as "catching" light particles is flawed.

For centuries, the dominant story of vision has been built on a simple, intuitive metaphor: *light travels from an object, enters the eye, and paints an image on the retina.* It's taught in schools, embedded in scientific illustrations, and repeated in everyday language—phrases like "light enters the eye" or "photons hitting the retina" reinforce the notion that sight is a passive process of catching projectiles launched from the outside world.

This model works as a basic teaching tool, but it begins to collapse under scrutiny. If vision were simply a matter of collecting incoming particles, we'd expect to encounter significant delays in perception, especially across vast distances. Yet even in real-world, non-astronomical conditions, our perception of events feels immediate—far faster than the travel times implied by the projectile model.

The problems deepen when we look at uniform perception. If photons were arriving from countless angles at slightly different times, there should be measurable distortions in our continuous sense of the world. Instead, our visual experience is remarkably stable, as if the whole scene is being resolved all at once rather than stitched together from a chaotic photon stream.

Low-light vision also poses a challenge. The photon model says that in dim conditions, fewer light particles are available, so perception should degrade in a predictable, granular way. In reality, human vision often maintains surprising coherence and pattern recognition, even when photon counts—according to standard calculations—should be negligible. The brain's ability to "fill in" is often cited as the explanation, but this is a patch, not a proof.

From the UED perspective, these inconsistencies vanish. Vision is not a process of gathering particles; it is the *alignment* of your observer-field with the ripple field of the observed object. No photon couriers are needed. Instead, the field's structured vacuum energy (SVE) reconfigures, and when alignment passes the Alignment Threshold Time (ATT), perception emerges—instantaneously, without spatial transit.

In this view, "seeing" is not about light traveling into you—it's about you *meeting* the field of what you are looking at. The illusion of photons as couriers is simply a byproduct of how we've been conditioned to think in terms of motion, cause, and effect, rather than coherence and alignment.

10.2 – The Eye as a Ripple Interface

Biological vision reinterpreted through ripple-field dynamics.

In the UED framework, the eye is not a passive container waiting for light particles to fall in—it is an *active interface* between the observer's biological structure and the surrounding ripple field. The retina, with its intricate arrangement of rods and cones, is not a photon-catching net. Instead, it is a field-sensitive matrix, finely tuned to detect and synchronize with variations in Structured Vacuum Energy (SVE).

From this perspective, visual perception begins not with a stream of incoming particles, but with a coherence event. The field of the observed object and the field of the observer establish an alignment relationship. The optic system is part of the observer-field, and when the local field alignment reaches the **Alignment Threshold Time (ATT)**, a visual event is triggered. This is not a transfer of physical matter; it's a reconfiguration of the field state that instantly updates the observer's experience.

The ATT here serves as the gatekeeper. Sub-threshold ripple activity—misaligned, unstable, or too weak—never registers as vision. Only when the alignment stabilizes at or beyond the threshold does the "image" emerge in awareness. This explains why perception can feel sudden, as if something "appears" without delay, even when conventional optics would predict a gradual build-up of clarity.

In this model, the image you see is not stored inside your eye or brain as a picture. What you experience is the *aligned field itself*—a living, dynamic relationship between you and the object's ripple structure. Your brain participates by decoding the coherence patterns into meaningful perception, but it does not hold the "scene" as a physical diagram.

The eye, therefore, is less like a camera and more like a tuning fork—it resonates with the field until the ripples between observer and object are in phase, and at that moment, perception is born. This shift from reception to resonance redefines vision as an act of participation, not passive reception.

10.3 – ATT and Visual Readiness

How visual perception begins before you think you "see."

In everyday life, we imagine that we start "seeing" the instant light reaches our eyes. UED dismantles this assumption. Long before you consciously register an object, ripple interactions are already unfolding in the background, operating below the Alignment Threshold Time (ATT). This pre-alignment phase is invisible to you—not because nothing is happening, but because the field has not yet stabilized into the coherent state required for perception.

When the alignment between your visual system's field and the object's ripple signature crosses the ATT, perception is triggered instantly. There is no gradual fade-in from darkness to image. Instead, it's a sudden switch: *one moment, no scene; the next, the world appears.* This is the field reconfiguring itself into a state your sensory network can interpret. The experience feels instantaneous because it is—a threshold crossing is not a transmission delay, but a state change.

Interestingly, vision feels continuous, yet under UED it is actually a rapid series of discrete field alignment events. The mind stitches these individual "successes" together into the seamless flow we call sight. This explains phenomena like motion blur—not as the smearing of particles, but as the brain smoothing multiple alignment snapshots into one coherent experience.

ATT in vision is influenced by both biological and neurological factors. On the biological side, retinal sensitivity, pupil adjustment, and even micro-saccadic eye movements affect how quickly field alignment is achieved. Neurologically, the brain's processing speed, attentional focus, and prior pattern recognition influence how efficiently these alignment events are registered.

In essence, vision is the mind's reading of field success events—each one a moment when ripple alignment surpasses ATT and collapses into conscious awareness. What you "see" is therefore not a stream of photons, but a rhythm of field victories, each one a triumph of coherence over noise.

10.4 – Vision as Mutual Alignment, Not One-Way Reception

You don't "pull in" an image—your field meets the world's field halfway.

The conventional picture of vision has always been one-directional: light leaves an object, travels through space, enters your eye, and paints an image on your retina. UED replaces this with a two-way handshake model. Your perceptual field is not a passive canvas—it is an active ripple structure, continuously seeking alignment with the ripple state of the external world.

In this framework, mutual readiness is the key. The object-field and the observer-field each have their own dynamic state, fluctuating with environmental conditions, attention, and intrinsic stability. The moment their ripple configurations reach coherence, perception is triggered. It is not a case of "light finally getting here," but of two oscillating systems suddenly snapping into phase.

This also explains why distant objects can appear instantly. When atmospheric clarity changes, when your eyes refocus, or when your attention shifts, the alignment threshold (ATT) between your field and the object's field can be crossed in a fraction of a second. From your perspective, the object "comes into view," but in reality, the change was in field readiness, not in photon travel time.

The process is remarkably similar to wireless syncing. Two devices can remain silent for hours, and then—in an instant—data flows as soon as the handshake conditions are met. The apparent "delay" is just waiting for the right key, not moving the data through a long pipeline. Likewise, in UED, vision is a sync event, not a delivery event.

By recognizing this, the last trace of the "light arrives" illusion dissolves. The perceived sequence of object → light → eye → brain is a human-friendly reconstruction, not a physical necessity. In truth, there is no photon journey to track—only the mutual settling of field states into a shared pattern of coherence.

10.5 – Vision Latency and the Illusion of Reaction Time

Delays in seeing are not due to light travel, but to alignment buildup.

In the mainstream view, reaction time is often treated as a chain of physical delays: light leaves an object, travels to your eyes, is converted into electrical impulses, then processed by the brain. This model assumes the delay begins with light "travel" and ends with brain interpretation. UED dissolves the first step entirely—light is not traveling to you, it is the *field* seeking alignment with your perceptual node. The true source of latency lies in the build-up of alignment until the Threshold (ATT) is crossed.

ATT variation explains why sometimes you notice things immediately and sometimes with a subtle lag. For example, when you walk into a dim room, the scene doesn't fade in gradually because photons are arriving—it appears suddenly because the field state between your vision system and the room's ripple structure crosses the coherence threshold. Until then, the world is effectively invisible to you, even though the ripple field is active.

This also explains why sudden visibility feels like a switch, not a fade. The shift from "nothing" to "clear view" is a binary perceptual event, because the field alignment is a tipping point phenomenon. Below ATT, no image forms in awareness; above ATT, the image exists in full perceptual fidelity.

Some species illustrate this principle dramatically. High-speed perception in certain animals—like birds of prey tracking insects mid-flight—comes from lower ATT requirements. Their visual systems can achieve ripple coherence with fewer cycles of field alignment, enabling perception to update more frequently. This gives them an almost slow-motion view of reality compared to humans, without breaking any laws of light "speed," because speed is never the limiting factor.

From a human performance perspective, perceptual speed training is not about "faster reception" but about tuning field sensitivity—reducing the cycles needed for coherence. Martial artists, elite drivers, and trained observers often develop this ability unconsciously. They are not catching photons faster; they are reaching perceptual readiness sooner.

10.6 – Implications for Biology and Technology

What changes when we stop treating vision as photon-catching.

Recasting vision as field alignment rather than photon capture rewrites the design principles of both biology and technology. If seeing is not the act of catching traveling particles, then all instruments and interpretations built on that premise are already working on an outdated model—they're just compensating for the wrong cause.

In technology, this means that imaging systems could be redesigned to work as field alignment detectors instead of light collectors. Cameras today rely on lenses and sensors to focus and trap photons, then translate them into digital signals. Under UED, the future could be devices that bypass the "collection" step entirely, tapping directly into the structured vacuum energy (SVE) to map ripple coherence states. Such devices would not be constrained by line-of-sight or lighting conditions—they would be able to register a coherent alignment event even in darkness or occlusion, because the field itself is the carrier of perceptual reality.

In biology, this reframing opens new ways to understand phenomena like hallucinations, mirages, and optical illusions—not as *errors of reception*, but as false alignments. Here, the visual field achieves temporary coherence with ripple patterns that do not correspond to the physical configuration we expect. The brain isn't "fooled by faulty light"—it is successfully aligning with a ripple state that simply doesn't match the one we think we're in. This also redefines visual disorders: some may not be sensory damage at all, but a chronic misalignment between internal and external field states.

The same principle suggests bio-augmentation possibilities. If vision is not tied to the eyes but to the nervous system's ability to align with field information, then in theory, "direct field vision" could be possible—extending visual awareness beyond the physical organ through artificial interfaces. Such technology could allow a human to perceive their surroundings without the use of eyes at all, by stimulating ripple alignment patterns directly in the neural field.

Finally, there is a philosophical shift: the universe is not "seen" as if it were a fixed display awaiting our gaze—it is co-experienced in the moment of alignment. You don't look at the world; you meet it halfway in a shared field event. Every act of seeing becomes a handshake with reality, where observer and observed are not separate points on a map but nodes in a living coherence network.

Thus this shows how UED vision theory opens both scientific and philosophical frontiers—turning vision from a passive act of reception into an active process of co-creation between biological systems and the field itself.

Chapter 11: Color Is Not Wavelength

It's Meaning

Psychological and ripple-level explanation of color perception

We've been told since school that *color is just wavelength*—a tidy equation linking frequency to hue, as if the universe were handing out shades like a paint chart in a hardware store. But that neat story leaves too much unexplained: why the same surface can look different under changing light, why two people can disagree on the same shade, why colors stir emotions or memories that physics alone can't account for.

From the UED perspective, color is not an inherent property of light at all. It is a meaningful alignment state—a specific ripple configuration between the observer and the field. What you "see" as red, blue, or gold is not traveling toward you; it's emerging in the moment of coherence between your sensory field and the structured vacuum's pattern.

In this chapter, we will bridge physics, psychology, and ripple mechanics to reframe color entirely—not as a fixed physical trait, but as the mind's code for a successful handshake with reality. By the time you finish, "wavelength" will feel like a footnote, and "meaning" will take center stage.

11.1 – The Wavelength Dogma

How the mainstream reduces color to a physical measurement.

The story every science textbook tells is seductively simple: light has a frequency, that frequency determines its photon energy, your retina catches those photons, and your brain dutifully translates that into a color. It's an elegant chain:

light frequency → photon energy → retinal stimulation → color perception.

In this view, color is nothing more than a label on the energy of a photon, a strict mathematical mapping between wavelength and hue. It's clean, measurable, and fits beautifully into the framework of classical optics and quantum theory.

But the cracks in this story are obvious the moment you step out of the lab and into real life. First, color constancy throws a wrench into the "wavelength = color" idea. A white shirt looks white under sunlight, under a warm incandescent bulb, and even under the cold bluish glare of LEDs—despite the fact that the spectrum of light hitting it is wildly different in each case. If color were just wavelength, the shirt should appear drastically different every time.

Second, human diversity in color perception undermines the claim of a universal mapping. Color blindness alters or erases distinctions that the wavelength model says should be fixed. Synesthetes experience colors linked to sounds, numbers, or even personalities—phenomena that wavelength alone cannot explain because no photon frequency is involved in those triggers at all.

Third, there's the elephant in the room: the emotional and symbolic dimensions of color. Red doesn't just "measure" at around 700 nm—it signals danger, passion, or power in human minds. Blue can soothe or chill. Gold evokes value and reverence. None of these associations arise from physics; they are meaning patterns woven into the cognitive fabric of the observer.

From a UED perspective, the problem with the wavelength dogma is that it mistakes a trigger condition for the experience itself. Wavelength is simply one possible cue for a particular ripple alignment between observer and field. Color is the *result* of that alignment—not the property of a photon.

11.2 – Ripple Signatures, Not Frequencies

In UED, color corresponds to the pattern of ripple alignment, not the frequency of a traveling wave.

If mainstream physics treats color as the output of a lightwave's frequency, UED throws that assumption out entirely. In this framework, color is not about how fast a wave "vibrates" through space—it's about the specific signature of ripple alignment in Structured Vacuum Energy (SVE).

SVE is not an empty void; it is a structured, still field capable of holding coherent ripple patterns. These patterns are not "waves traveling from here to there" but stable states of alignment between an object-field and an observer-field. When your visual system tunes into such a state, your brain translates it into what you call "color."

This explains a mystery the wavelength model can't touch: why two completely different sources—say, a sunset sky and the glow from a sodium vapor lamp—can produce the exact same perceived color, even if their physical spectra differ dramatically. From a UED standpoint, this happens because the two events share a matching ripple alignment signature, not because they emit photons of identical frequency.

Furthermore, a ripple state can shift in perception without any measurable change in its frequency. Think of watching a painting in daylight, then at dusk. The measurable spectra of light from the painting may barely change, but the surrounding field context alters the alignment state between you and the image—your experience of its colors deepens, softens, or becomes more vivid. This is not "light changing"—it's the *field relationship* changing.

In this way, frequency is just one potential side-effect of a ripple state, not its defining essence. The core reality is the alignment code—a specific pattern of tensions and releases within the SVE field that the biological system interprets as a particular color experience.

11.3 – Color as a Meaning Assignment

The mind does not "see" color—it assigns meaning to field states.

In the UED view, color is not a property of light, nor is it something "out there" waiting to be collected by the eyes. Color is a mental assignment—a label your mind applies when it registers a particular ripple alignment success event with the field.

When your eye–field system locks into a specific SVE pattern, your brain doesn't just *register* it mechanically; it *interprets* it. That interpretation comes wrapped in memory, emotion, language, and expectation. The result is what you call "red," "blue," or "emerald green."

Because of this, color perception is context-dependent by design. The same ripple alignment may feel entirely different depending on surrounding conditions—lighting, background objects, or even your emotional state. A white shirt under midday sunlight has a different "feel" than the same shirt under a streetlamp at night, not because the photons changed in any meaningful way, but because the field conditions have altered the mind's interpretation of that alignment.

Cultural and linguistic frameworks amplify this effect. Some languages do not separate "blue" and "green" as distinct categories—speakers simply interpret a range of ripple signatures as one conceptual color. In other cultures, there are dozens of distinct names for what English lumps together as "white" or "red." This isn't a limitation of the eye—it's evidence that color lives in the mind–field partnership, not in any fixed spectrum.

Even personal experience can shift meaning assignments. An artist, a soldier, and a botanist might each look at the same shade of green and assign different mental weight to it—one sees vibrancy and life, another sees camouflage, another sees a particular plant species. The ripple signature is the same; the meaning it carries is entirely dependent on who aligns with it and how their mind chooses to interpret it.

11.4 – The Emotional Layer of Color

Why colors carry mood, symbolism, and psychological effect.

In the UED framework, ripple alignment does more than trigger a sensory registration—it cascades into your emotional field. The same structured vacuum energy (SVE) alignment that informs the mind "this is green" can simultaneously nudge your emotional state toward calm, alertness, excitement, or even discomfort.

This is why color is never just visual data. It is an *experience node*, activating sensory, cognitive, and emotional layers at once. For example, "red" is often linked to urgency, heat, danger, or passion—not because red photons inherently "carry" those qualities, but because red's ripple signature historically coincides with survival-relevant events (blood, fire, warning signals). Over generations, these associations become deeply embedded in the way minds assign meaning to that alignment.

Similarly, "blue" tends to evoke calm, depth, or trust—not because blue light has a sedative chemical effect, but because its ripple alignment is reminiscent of large, stable, life-supporting fields like the ocean or clear sky. That said, these associations are widespread but not universal—cultural narratives and personal experience can override or reshape them. In some cultures, white is the color of purity; in others, it is the color of mourning.

This emotional coupling explains why marketing, art, and ritual practices use color so deliberately. A brand's "trustworthy" blue, a temple's "sacred" gold, or a protest movement's "urgent" red banner—these choices are not arbitrary design elements. They are field-based emotional triggers, chosen to induce specific alignments between the collective emotional field of the audience and the intended message or mood.

In UED terms, when you "feel" a color, you are not reacting to a beam of photons—you are resonating with a multi-layered field state in which sensory perception, memory, and emotion converge. This also explains why certain colors feel "draining" or "energizing" without any measurable change in light intensity—your emotional field is responding to the ripple signature's alignment, not to raw physics.

11.5 - ATT and Color Stability

How Action Trigger Threshold (ATT) influences color perception.

In UED, color stability is not a property of light—it's a property of alignment maintenance. Once your visual field crosses the Action Trigger Threshold (ATT) for a given ripple signature, the perception of that "color" locks in. As long as this alignment is maintained above ATT, the color appears stable—even if the illumination changes, shadows fall, or the physical conditions shift. This is why an apple still looks "red" under both midday sunlight and indoor lamplight: the ripple signature remains in the same alignment zone, and the mind continues to assign the same color meaning.

When alignment dips near or below ATT, perceptual instability emerges. This is when sudden hue shifts can occur, even though the physical environment hasn't changed in a way that classical optics would predict. The viral internet debate over "*the dress*"—whether it was blue/black or white/gold—is a textbook example. Different observers' visual systems locked into different alignment states, crossing ATT at different points, creating mutually exclusive yet equally "real" perceptions.

ATT also explains color adaptation when moving from bright sunlight to dim light. In bright conditions, many ripple signatures are above ATT simultaneously, allowing rich color differentiation. In low light, fewer signatures cross ATT, forcing the visual system to work with a reduced palette. The brain fills the gaps by relying on memory, expectation, and contrast, which is why colors may appear muted or shifted until alignment stabilizes again.

Even short-term anomalies—like momentary flashes of false color when looking at bright objects after stepping into darkness—are ATT artifacts. These are not "afterimages" in the traditional sense, but the visual field re-stabilizing its ripple alignment under new lighting conditions.

11.6 – Color Without Light

Why you can "see" color in dreams, hallucinations, or closed-eye visions.

One of the clearest demonstrations that color is not an external property of light is the fact that you can vividly experience it without any photons entering your eyes. In dreams, hallucinations, or even simple "mind's-eye" imagination, color can feel just as real as it does in waking life under full daylight. From the UED perspective, these experiences occur when the mind–field system achieves an internal alignment state that matches a stored ripple signature—the very same alignment your visual system would produce if the corresponding real-world object were in front of you.

In this scenario, there is no external stimulus, no SVE interaction with the optic interface, and no photon-triggered retinal activity. Instead, the mind is recalling or synthesizing a coherence pattern from memory, imagination, or other neural processes. The moment the brain's internal field matches the stored signature, the perceptual experience of that color emerges—regardless of whether your eyes are open or closed.

This explains why:

- **Dreams** can have richly saturated landscapes and faces with lifelike skin tones—your mind is not "seeing" light but recreating alignment states.
- **Hallucinations** (from illness, substances, or sensory deprivation) can introduce colors that have no physical origin in your environment but still feel absolutely present.
- **Closed-eye visuals**—such as swirling colors when pressing on the eyelids—are ripple disturbances generated internally, crossing ATT without any external field involvement.

The phenomenon extends into synesthesia, where certain sounds, words, or even numbers trigger consistent color experiences. Here, the brain's sensory networks link ripple patterns from one domain (e.g., sound) to stored visual alignment signatures, generating color without light.

In UED, this means color is not "out there" in the world—it is an interpretive state of the mind–field system. External light is simply one way of reaching that alignment; internal neural processes are another. This insight dissolves the rigid connection between color and wavelength, showing instead that color exists anywhere alignment meaning is invoked—whether through perception, memory, or imagination.

11.7 – Implications for Science, Art, and Technology

The new possibilities when color is treated as meaning, not wavelength.

If color is redefined from a property of light to a meaningful ripple alignment state, the consequences ripple far beyond the scope of traditional optics. It shifts the way we approach art, science, technology, and even philosophy—because it proves that perception is not about passively receiving reality, but about actively participating in its formation.

Art – Beyond Pigments and Screens

Artists would no longer be bound by the limitations of physical pigments or digital RGB screens. Instead of relying on paint, ink, or pixels to reflect or emit certain wavelengths, creators could design direct alignment triggers that evoke specific color experiences in the observer's mind–field system. Imagine an installation that, without projecting light, can make you "see" a deep oceanic blue or fiery crimson simply by tuning your internal alignment state. Art could bypass the eye entirely and communicate pure perceptual meaning.

Science – Reading the Invisible Through Color

In UED terms, "color" could become a diagnostic tool for detecting field states that are invisible to conventional sensors. Instead of measuring wavelength, instruments could be calibrated to translate ripple signatures into perceptual color outputs—allowing scientists to "see" electromagnetic, gravitational, or biological field changes without requiring visible light. For example, a medical scanner might represent tissue health as specific color alignments in the observer's mind, bypassing conventional imaging entirely.

Technology – Mood-Aligned Environmental Control

If color is an emotional and alignment state, not just a visual stimulus, environments could be tuned to align people's fields for desired effects—calmness, focus, creativity—without relying on lighting changes. Future devices could modulate ripple fields in a room so that individuals *feel* the presence of a color experience even with their eyes closed. This could revolutionize therapy, workspace design, and entertainment.

Philosophy – Color as Proof of Co-Creation

The UED model turns color into a case study for perception as participation. You don't just "see" a color; you meet it halfway, completing an alignment handshake between your mind–field and the universe's field. This dissolves the boundary between "out there" and "in here," showing that all perception—color included—is a co-authored reality.

Closing Reflection – "The Universe Never Shows Its True Colors"

Color, as we have been taught to believe, is something "out there"—a pigment, a frequency, a measurable slice of the electromagnetic spectrum. But in the UED framework, this notion dissolves. The universe itself holds no color cards. It does not "wear" blue skies or "paint" sunsets. Instead, what we call color is the mind's poetic handshake with the field—a moment when your internal alignment meets an external ripple state, and together they co-create an experience we label as red, green, violet, or gold.

In that handshake, there is no ownership. The field doesn't own the hue, and neither do you. Each shade is a shared invention between you and reality, an agreement born of alignment. Change the alignment, and the meaning shifts. The same field state that feels warm and inviting to one observer may feel cold and distant to another, because color exists only in the space where perception and existence overlap.

In this sense, the phrase "true colors" becomes misleading. The universe never reveals a fixed palette—it only offers ripple possibilities. The mind, with its history, biology, and intent, selects which ones to experience. That's why sunsets can be nostalgic, ceremonial robes can be sacred, and a certain green can remind you of your grandmother's garden. None of these colors are stored in photons; they are stored in the living connection between you and the world.

When you realize this, you stop looking for the "real" color of anything. You start to see that color is not truth—it is relationship.

And in that relationship, the universe never stops changing its mind, because it never had a single hue to begin with.

Chapter 12: Reflection, Refraction, Shadow

Field Events, Not Beam Tricks

Classical optics reinterpreted via SVE ripple mechanics

For centuries, we've been told that light behaves like a dutiful courier—racing from one place to another, bouncing off mirrors, bending through glass, or being blocked to cast shadows. The diagrams in textbooks make it look almost mechanical: neat arrows showing a beam's journey, angles calculated as if light were a ball ricocheting off a wall.

But what if none of that is actually happening?

In the Universal Energy Dynamics (UED) framework, reflection, refraction, and shadow are not the result of photons traveling, colliding, or being stopped. They are instantaneous field state transitions within the Structured Vacuum Energy (SVE) that pervades reality. Surfaces don't send light back to you—they reshape the field so that alignment between you and the object reconfigures. A shadow isn't light "failing to arrive"—it's a region where alignment never happens. Refraction isn't light slowing down—it's the ripple field finding a new orientation to maintain coherence across boundaries.

In this chapter, we will dismantle the courier myth and reveal that these classical optical effects are not time-delayed beam tricks, but immediate responses of a living field. The world doesn't wait for light to reach you—it reconfigures itself with you.

12.1 – The Beam Illusion in Classical Optics

For generations, optics has been taught through a simple, almost comforting narrative: light behaves like a well-trained traveler. It moves from one place to another, bounces when it meets a mirror, slows down and changes course when it passes through glass, and stops entirely when blocked—leaving behind a shadow. This picture is drilled into us with neat ray diagrams and a language that makes light sound like a stream of tiny couriers or oscillating waves on a journey.

In the mainstream interpretations, this plays out as:

- **Reflection** – A photon is said to strike a surface and "bounce" off, obeying the law of equal angles.
- **Refraction** – The photon supposedly slows down when entering a denser medium, causing it to bend toward the normal, then speeds up again when leaving.
- **Shadows** – Where an object blocks light, no photons make it to that region, and so it remains dark.

While elegant on paper, this narrative introduces deep conceptual limits and contradictions:

- **Speed Paradox** – If photons physically travel, how do we account for the apparent instant appearance of distant reflections or the uniformity of illumination over vast areas?
- **Simultaneity Conflicts** – The beam model cannot cleanly explain why, in certain optical setups, changes in alignment

seem to produce immediate effects without measurable delay.

- **Sharpness and Coherence** – Perfectly crisp shadows or reflections would require an impossible precision of countless individual photon trajectories, yet they remain stable under wildly varying conditions.

From a UED perspective, these contradictions vanish when we drop the courier metaphor entirely. Reflection, refraction, and shadow are not the result of photon traffic. They are local ripple realignments within Structured Vacuum Energy (SVE). Surfaces, boundaries, and obstacles do not send or block photons—they alter the way the field is structured between observer and environment. Once that structure changes, perception updates instantly upon crossing the Alignment Trigger Threshold (ATT).

The beam illusion survives only because it's easy to draw. The field truth survives because it's how reality actually works.

12.2 – Reflection as Field Realignment

In the UED framework, a reflection is not a parcel of light being "returned" to you after a journey. It is the instantaneous co-creation of an image within your perceptual field—arising when the ripple state of your own field aligns with that of the object through the mediation of a reflective surface.

A mirror does not "throw" the object's photons back at you like tiny rubber balls. Instead, it acts as a ripple alignment amplifier, preserving the structural coherence of the field between you and the object. When the surface is smooth enough at the microscopic level, it avoids introducing disruptive micro-ripples, allowing the field state to remain coherent and complete the alignment handshake between object and observer. The result is what we call a reflection—but in reality, it's not a "return trip" of anything; it's a simultaneous alignment event.

Because this is a field process, surface smoothness is less about geometric perfection in the classical sense and more about minimizing ripple distortion. A highly polished mirror creates a clean, undisturbed alignment pathway, while a rough surface scatters the ripple coherence into multiple, incomplete alignments—producing diffuse reflections or none at all.

The instantaneity of reflection under UED follows naturally. There is no need to account for the "out-and-back" travel time of photons. The moment the ripple alignment conditions are met—when the Alignment Trigger Threshold (ATT) is crossed—your perceptual field updates with the image. This is why reflections feel immediate, whether you're a centimeter away from the mirror or many meters across a smooth lake.

Clarity of reflection, then, becomes a direct function of ATT stability:

- If ATT is low and conditions are steady, the reflection appears crisp and solid.
- If ATT is unstable due to turbulence, vibration, or field noise, the image appears wavy, blurry, or intermittently broken.

From this view, a reflection is not a replay of visual data but a live co-alignment—a mutual field state where you and the object meet in the "middle" of reality, orchestrated by the properties of the reflective surface.

12.3 – Refraction as Ripple Path Reconfiguration

In mainstream optics, refraction is framed through Snell's law, where light changes speed when it enters a medium of different optical density, causing its path to bend. While this model is mathematically serviceable for many engineering purposes, it relies on the beam travel assumption—that light is a moving entity whose trajectory can be "deflected" by a change in velocity. In the UED framework, this interpretation is both unnecessary and incomplete.

In Universal Energy Dynamics (UED), refraction is not about a photon slowing down or a wave compressing—it is about ripple path reconfiguration at the interface between two different field states. When a ripple encounters a medium boundary (e.g., air to water, vacuum to glass), the Structured Vacuum Energy (SVE) field at that boundary exhibits a change in ripple tension. This change is not a mechanical drag; it is a shift in the alignment gradient—the pattern by which the object's field and the observer's field maintain coherence through the medium.

The so-called "bending" is, therefore, not a literal curve in the trajectory of a physical particle, but a reorientation of the alignment pathway. The ripple state reorganizes itself to preserve coherence across the interface, and the mind interprets this reorganized alignment as a displaced image. In simpler terms: the scene didn't bend—your alignment with it shifted.

This perspective elegantly resolves certain inconsistencies in classical optics:

Speed-change paradox: Experiments show light's *measured* transit time through transparent media can be affected by boundary effects and coherence conditions, contradicting a simple speed-change model. In UED, no "slowing" occurs—only a recalibration of alignment at the boundary.

Mirages: Conventional models require complex ray-bending diagrams to explain inverted and displaced images in air layers of varying temperature. UED treats this as a gradual shift in ripple tension, producing a smooth reconfiguration of alignment without any "wave steering."

Atmospheric bending: The Sun appears above the horizon even when it is geometrically below it. In UED, the field alignment between the Sun and observer adjusts continuously through layers of varying ripple tension in the atmosphere—there is no need to imagine beams physically bending for kilometers.

Under this model, Snell's law becomes a description of boundary-induced reorientation ratios rather than a statement about light's velocity. The equation still works numerically in many contexts, but its physical meaning shifts from "slowing and turning" to "re-aligning to maintain coherence."

In essence, refraction in UED is the field's adaptive handshake—a recalibration to keep perception whole, even when the medium changes.

12.4 – Shadow as Alignment Absence, Not Light Blocking

In the beam-based paradigm, a shadow is described as the region where photons from a light source are physically blocked by an intervening object. This explanation assumes that light is a tangible courier, and that its absence produces darkness. But in UED, the concept of a shadow emerges from an entirely different principle—not as a shortage of particles, but as a gap in alignment.

A shadow zone is simply a region in which the field coherence between the source and the observer is disrupted by the presence of an intervening object. The object interrupts the ripple handshake—the mutual alignment between source field and observer field—by inserting its own ripple signature into the interaction. This doesn't mean the field is "empty" or "unlit." It means that the original alignment channel has been overwritten by the object's field, preventing the direct perception of the source in that area.

This model also explains shadow edges far more naturally than photon-blocking does:

- Edge blur and penumbra occur when the field alignment is only partially disrupted. The surrounding space still leaks some coherence from the source–observer channel, producing a gradient of alignment strength.
- In beam theory, this requires imagining partial photon coverage and overlapping beam cones; in UED, it's simply a partial ripple overlay where two alignment states compete in the same zone.

Crucially, UED predicts and explains shadows beyond the visible spectrum without requiring a "different set of particles" for each band:

- **Infrared and radio**: An object's field can disrupt alignment in these bands even if it is transparent to visible light, producing a "shadow" in that field channel.
- **Gravitational or other subtle fields**: A massive body can produce a shadow-like absence in certain gravitational-wave or structured vacuum alignments, even though nothing is physically "blocking" anything.

From this perspective, a shadow is not the absence of light—it is the presence of a different alignment state. The dark patch you see is not an empty space; it is a field region whose handshake with the source has been replaced by the intervening object's own ripple imprint.

Under this view, darkness is not emptiness—it is an alternative field conversation happening in the same space.

12.5 – UED Unification of Reflection, Refraction, and Shadow

All three classical "light phenomena" are not separate tricks of a traveling beam, but different outcomes of the same fundamental process — ripple-field alignment and its modulation by local conditions.

Single principle: Whether light appears to bounce, bend, or vanish, what is actually changing is how the Structured Vacuum Energy (SVE) in a given region supports or disrupts alignment between the observer and the source.

- Reflection occurs when the alignment pathway is preserved via a surface that reorients ripple states toward the observer.
- Refraction occurs when the alignment pathway is preserved but reconfigured to account for ripple tension changes across media.
- Shadow occurs when the alignment pathway is disrupted or blocked from forming.

Common field mechanism: All three depend on the same variables:

- Ripple tension (local SVE strain state)
- Alignment Threshold Time (ATT) (minimum time for perception trigger)
- Field memory (persistence of prior alignment patterns) The difference lies only in *how* these variables are influenced by

the object or medium in the field.

Why they feel different: Classical optics describes each as a separate event because they appear visually distinct — a mirrored image, a bent straw in water, a dark outline on the ground. UED shows these are different ripple solutions to the same alignment equation:

- In reflection, the equation resolves by inverting or redirecting the ripple map.
- In refraction, it resolves by remapping without breaking the path.
- In shadow, it resolves by failing to meet ATT within the blocked zone.

Instantaneity explained: Because no photons travel from A to B and back, all three occur at field speed — the maximum reconfiguration rate of the SVE (our redefined *c*). This means the visual outcome is updated instantly across the alignment network, limited only by ATT.

Why this matters: Understanding them as one process simplifies optics dramatically:

- One foundational law of alignment replaces three separate beam-based explanations.
- Engineering applications (mirrors, lenses, imaging systems) can be designed by direct field control instead of beam manipulation.
- It becomes possible to create "shadow," "reflection," or "refraction" in non-visible fields (gravity, magnetism, bio-fields) using the same principles.

12.6 – ATT, Field Memory, and Optical Persistence

What we call "stability" or "persistence" in optical effects is not due to lingering light, but to the interplay between Alignment Threshold Time (ATT) and the SVE's field memory. Together, they dictate how long a reflection holds steady, how refraction maintains an image, and how shadow boundaries remain sharp or blur.

1. ATT as the stabilizer

- In UED, ATT is the minimum temporal coherence required between the observer's field and the source's field for perception to "trigger."
- In reflection, if ATT is consistently met, the image appears stable on the mirror; if it is only intermittently met (e.g., moving water surface), the reflection shimmers.
- In refraction, ATT stability ensures a straw looks bent in water without flickering or distortion — unless turbulence in the medium pushes the alignment under threshold.
- In shadow, ATT determines how sharply the boundary is perceived; unstable ATT yields a soft or moving penumbra.

2. Field memory — the invisible afterimage

- The Structured Vacuum Energy field retains residual alignment patterns for a short time after the source or alignment conditions change — much like ripples in water after a stone is removed.
- This "field memory" explains persistence of vision (afterimages) without invoking residual photon bombardment or retinal fatigue alone.
- The lingering glow after staring at a bright object, or the

phantom image of a moving fan blade, is the mind continuing to read the fading alignment pattern before it fully dissipates.

3. No residual photon flow required

- In beam-based optics, persistence is awkwardly explained as "light still arriving" or "the retina still excited."
- UED removes the need for traveling photons entirely — persistence is simply the time it takes for the field to lose coherence below ATT.
- This makes optical persistence a universal phenomenon across all perceptual channels, not just vision (e.g., "ghost notes" in sound, lingering scents).

4. Fast motion and distortions

- Fast-moving objects sometimes appear stretched, delayed, or partially transparent — not because the eye "can't keep up," but because rapid field reconfiguration breaks ATT continuity.
- The visual system "stitches" incomplete alignment events together, producing motion blur or lag.
- Animals with shorter ATT (e.g., birds of prey) perceive far less distortion because their fields can trigger perception from smaller coherence windows.

5. Implications

- **Biology:** Perceptual training (e.g., sports, hunting) could focus on tuning ATT sensitivity rather than merely "improving reaction time."
- **Technology:** Cameras and sensors could mimic ATT-field

memory mechanics for smoother image capture without frame lag.

- **Philosophy:** What we call "seeing" is not a constant stream but a mosaic of discrete alignment events, glued together by the field's capacity to hold coherence briefly beyond the moment of exact alignment.

12.7 – Practical & Technological Implications

Once reflection, refraction, and shadow are seen as field events instead of photon games, the design principles for optical systems shift radically—from *light gathering* to *field alignment*. This opens new avenues for engineering, imaging, and even biological augmentation.

1. Rethinking optical design

- In the mainstream approach, mirrors, lenses, and sensors are optimized for capturing, directing, or focusing *streams of light*.
- Under UED, the goal changes: maximize and stabilize field coherence between source, surface, and observer.
- Mirror quality becomes less about perfect surface polish alone and more about surface–field compatibility—the ability of a surface to mediate precise ripple realignment.
- Lenses could be engineered not to "bend" light, but to facilitate smoother field state transitions across medium boundaries, eliminating chromatic aberration and dispersion effects at their root.

2. Imaging systems that "align" rather than "collect"

- A UED-based camera would not rely solely on photons

entering a sensor array—it would tune itself to the SVE ripple signature of the target field.

- This means imaging could work in total darkness, dense fog, or underwater without traditional illumination, as long as the system can handshake with the target's ripple state.
- Such devices could read "optical" information from environments where no visible light exists, bypassing wavelength limitations entirely.

3. Real-time, medium-independent vision

- Since alignment is a field event, not a beam dependency, perception (human or machine) could be engineered to work seamlessly across air, water, glass, or even dense materials.
- This makes possible true all-terrain vision systems for robotics, aviation, and medical imaging, where traditional optics fail due to scattering or absorption.
- In biology, this could inspire bio-augmentation—adding sensory organs or implants that detect field states directly, creating vision that is immune to glare, darkness, or fog.

4. Shadow analysis as a mapping tool

- In UED, shadows are not the "absence of light" but regions of disrupted field alignment.
- By studying the shape, sharpness, and movement of these alignment gaps, it becomes possible to map field disturbances, detect hidden objects, or even identify material properties.
- This could lead to passive scanning systems where no active illumination is used—only the analysis of how field coherence patterns are interrupted.

5. Cross-disciplinary potential

- **Architecture:** Design spaces for optimal field alignment to enhance natural visibility and comfort without relying on high illumination levels.
- **Medical Imaging:** Develop non-invasive diagnostic tools that detect ripple misalignments in tissue, offering early disease detection without radiation.
- **Astronomy:** Replace photon-count-based telescopes with field alignment arrays, eliminating the constraints of cosmic "light travel time" and enabling near-instant field reads from distant bodies.

Closing Reflection – "No Bounce, No Bend, No Block"

The story you were taught about light is simple—perhaps too simple. It's the tale of a restless courier, darting across space, bouncing off mirrors, bending at boundaries, and disappearing behind obstacles. It's a story of travel, of particles on the move or waves in transit. But the UED view strips away the courier and leaves only the conversation.

Reflection is not a beam rebounding—it's a field handshake, a perfectly timed ripple alignment between observer and object, mediated by whatever surface stands between. Refraction is not light slowing down or changing direction—it's the field reconfiguring its pathways to maintain coherence across different ripple environments. And shadow is not the absence of a beam—it's the silence of alignment, a gap in the conversation where no handshake occurs.

In this view, optics is no longer the physics of travel—it is the art of readiness and relationship. The surfaces, spaces, and media you encounter are not obstacles or lenses for a traveling particle—they are participants in a constant ripple dialogue with your field. The crispness of a mirror, the bend of a straw in a glass of water, the soft penumbra at the edge of a tree's shadow—all of it is the field's language, speaking to you in the grammar of alignment and coherence.

When you drop the idea of "light as a courier," you also drop the baggage of delay, distance, and delivery. What's left is immediacy—events that happen because the conditions for them are met, not because something made the trip to your eye. Optics becomes not about chasing light, but about meeting the field halfway. And once you see that, you realize the truth in the phrase: No bounce. No bend. No block. Only the field, adjusting itself to keep the conversation alive.

Chapter 13: You Don't See the World

You Align With It

Perception as real-time field reconfiguration

We grow up believing the world streams into us—that light pours into our eyes, sound flows into our ears, and reality is delivered like a package to the brain. But under UED, that picture collapses. You never actually **receive** the world. You **meet** it.

Perception is not a shipment of photons and sound waves arriving at your doorstep—it is a handshake, a momentary coherence between your field and the field of what you call "out there." The tree you "see" is not sending you its image; you and the tree's field are aligning in the Structured Vacuum Energy.

In this chapter, we strip away the comfort of passive observation and replace it with the truth of active participation. You will see why perception is not a one-way pipeline but a two-way dialogue—instant, intimate, and constantly renegotiated. Once you understand this, you'll never again think you are "looking at" the world. You'll know you're **co-creating the moment you see.**

13.1 – The Myth of Passive Perception

For centuries, the dominant story of perception has been a comfortable one: your senses are windows, reality streams in, and your brain acts as the control center, stitching together the incoming data into the "picture" you experience. It's a neat, linear chain—light hits the retina, sound vibrates the eardrum, the brain processes the signals, and voilà—perception.

But this pipeline model is riddled with cracks when examined closely.

- **Delay:** If perception were a purely sequential process of reception and processing, there should be a noticeable lag in everything you experience. Yet your reactions, even in high-speed scenarios, are often faster than this model allows.
- **Selective Attention:** You do not "see" or "hear" everything in your environment—only what aligns with your current state of readiness. This is less about physical reception and more about what your system chooses to lock onto.
- **Illusions:** The very existence of optical and auditory illusions shows that perception is not a simple replay of external reality. If the pipeline were perfect, the data would be consistent, and no illusions would occur.
- **Contextual Shifts:** The same stimulus can be perceived differently depending on the surrounding conditions, emotional state, or expectation—impossible to reconcile with a strict data-in/data-out model.

From the UED perspective, these anomalies are not exceptions—they are the rule. You don't passively receive the world; you align with it in real time. Perception emerges as a *coherence event*—a handshake between your field and the environment's field—mediated by Structured Vacuum Energy (SVE).

In this view, "seeing" a tree is not about photons leaving the tree, traveling across space, and entering your eyes. It's about your field reaching an alignment threshold with the tree's field, producing a stable perceptual moment. The clarity, timing, and even emotional tone of that moment are determined by the quality of the alignment—not the volume of "data" received.

This shift—from perception as passive reception to perception as active alignment—changes everything. It means your senses are not open windows but dynamic tuning instruments, constantly adjusting to match the world's field state in microbursts of coherence.

13.2 – The Perception Handshake

Perception, in the UED sense, is not a one-way broadcast—it's a mutual agreement between your field and the environment's field. This is why we call it the *perception handshake*. Just like two devices on a wireless network cannot exchange data until they negotiate a compatible channel, your sensory system cannot "lock onto" reality until both sides reach a state of coherence.

Here's how it unfolds:

Seeking Coherence

Your sensory field—integrated across the eyes, ears, skin, and beyond—is constantly scanning for stable patterns in Structured Vacuum Energy (SVE). These patterns are not "signals" traveling through space; they are existing ripple configurations in the universal field. Your body's field extends into this environment, searching for states that can be brought into alignment.

Conditions for a Successful Handshake

For perception to click into place, three key criteria must be met:

- **Frequency Compatibility:** Your internal ripple dynamics must be able to resonate with the environmental ripple pattern. If the frequency gap is too wide, perception fails or distorts.
- **Ripple Stability:** The external field must hold its pattern long enough for your system to synchronize—unstable or

chaotic fields feel "blurry" or "flickering."

- **ATT Readiness:** Your Action Trigger Threshold (ATT) must be primed. If your readiness is low—due to distraction, fatigue, or competing stimuli—alignment either lags or never occurs, even if the external field is stable.

Instant Perception—No Travel Time

Once these conditions are satisfied, the handshake is immediate. There is no waiting for photons to cross a distance and be decoded; the shift happens at the point of alignment. That's why certain perceptions—like noticing a face in a crowd—can feel instantaneous, almost like they appear out of nowhere. It's not magic; it's the physics of field synchronization.

This handshake process also explains why perception can be lost and regained in bursts. When alignment breaks—through movement, obstruction, or field noise—the handshake fails, and the percept vanishes. Re-alignment re-establishes the handshake, and perception "returns" as if it had been waiting just out of reach.

In UED, perception is thus a living negotiation, not a pipeline delivery. Your field and the world's field are continuously adjusting to each other, producing reality as a series of micro-agreements—rapid, dynamic, and co-created.

13.3 – Sensory Systems as Alignment Nodes

In the UED framework, your sensory organs are not passive receivers of incoming data—they are alignment nodes, each tuned to a particular subset of ripple states in the Structured Vacuum Energy (SVE) field. Instead of "collecting" signals, they translate the patterns of local ripple coherence into forms your mind can integrate into experience.

1. Biological Ripple Translators

- Eyes are tuned to ripple patterns in the visual spectrum, not because photons "hit" them, but because their structure—retina, cones, rods—is optimized for locking onto certain SVE configurations that correspond to what we interpret as brightness, contrast, and color.
- Ears translate air-pressure ripple patterns into neural ripples, allowing the mind to align with vibrational events across space.
- Skin maps tactile ripple states—pressure, temperature, texture—into embodied awareness, allowing alignment with the physical environment's stability or disturbance.
- Nose is a chemical-ripple interface, detecting molecular field patterns and transforming them into olfactory experiences.
- Tongue works similarly, translating ripple-encoded molecular states into taste perception.

Each of these nodes works not by *catching traveling particles*, but by entering into resonance with the field's local configuration and relaying that alignment state to the mind.

2. Mapping Ripple Patterns into Experience

Every sensory node sends its translation not as raw physics, but as meaning-ready alignment states. The mind then weaves these into a coherent perceptual field. This is why what you "see" is already colored, shaped, and interpreted the moment you become aware of it—you are not processing raw data, but *field-aligned meaning*.

3. Reallocation of Alignment Bandwidth

When one sense is lost—such as vision in blindness—the common explanation is that other senses "compensate." UED reframes this: there is no *compensation*, only bandwidth reallocation. The total alignment capacity of your mind–body system is freed from one channel and redistributed among the remaining alignment nodes.

- This explains why blind individuals often have heightened hearing or touch—not because those systems grew stronger biologically, but because the alignment system now channels more field resources into them.
- Similarly, sensory deprivation in one modality can temporarily sharpen perception in another because alignment energy is no longer split across competing nodes.

In this view, perception is not about isolated senses feeding the brain—it's about a distributed network of alignment translators working in concert, each mapping the universal field into your unique, lived reality.

13.4 – ATT and Perceptual Resolution

In UED, Alignment Threshold Time (ATT) is the *temporal minimum* required for your field and the environment's field to achieve a stable alignment event. Only when ripple coherence crosses this threshold does perception "lock in" and become consciously available. Until then, the alignment is incomplete—like a handshake with fingers still hovering in the air.

1. Minimal Alignment for Awareness

- ATT defines the *point of readiness*: the moment when the sensory alignment nodes and the mind's integration system have stabilized the ripple handshake enough to trigger experience.
- If alignment builds slowly (e.g., watching a distant ship emerge from the fog), perception feels gradual. If it builds fast (e.g., a sudden face appearing in the dark), the event

snaps into awareness.

- In both cases, perception doesn't *fade in* continuously—it crosses a binary perceptual switch once ATT is met. This is why the world sometimes feels like it appears in chunks rather than in a seamless stream.

2. The "Pop" Effect in Awareness

- Jump scares in movies work by keeping ripple alignment just below ATT and then forcing a sudden coherence event—triggering shock, startle, and emotional spike in a single instant.
- Sudden recognition (e.g., seeing a friend in a crowd after scanning faces) is also an ATT crossing—the field pattern matches stored memory alignment, pushing perception over the threshold.
- This explains why you can look *at* something without really *seeing* it until ATT is satisfied.

3. Variations in ATT

ATT is not fixed—it varies:

- **Across individuals:** Some people have naturally lower ATT (faster perceptual readiness), giving them quicker reaction times and better situational awareness. Others require longer stability, which can slow perception but improve detail accuracy.
- **Across species:** Predatory birds or certain insects operate with drastically reduced ATT, allowing them to track high-speed movements that humans perceive as a blur.
- **Across mental states:** Fatigue, stress, meditation, or heightened alertness all modulate ATT.

    - Under stress, ATT may shorten for survival-relevant events (e.g., detecting a threat) but lengthen for non-urgent inputs.
    - In deep focus or meditation, ATT can lengthen, allowing perception of subtler, slower ripple changes.

4. Perceptual Resolution as a Function of ATT

A lower ATT improves temporal resolution—catching fleeting events—while a slightly longer ATT can enhance spatial or detail resolution, as the system has more time to integrate the field state before triggering awareness. This creates a trade-off between speed and richness of perception, determined by context and need.

In this way, ATT becomes the gatekeeper of perceptual reality—deciding not what exists, but what *appears* to exist for you at any given moment. The world you "see" is always filtered through this timing threshold, making perception an act of readiness, not mere reception.

13.5 – The Illusion of a Stable World

One of the deepest perceptual tricks your mind plays on you is convincing you that you live in a *continuous*, unbroken reality. In truth, according to UED, what you experience as a stable world is the result of rapid, successive field alignments—each one a discrete handshake between your field and the environment's field.

1. Perception as a Frame Sequence

- Every moment you are not "receiving" a stream of reality—you are aligning, disengaging, and re-aligning, over and over.
- These micro-alignment events occur at such high frequency that they merge into the illusion of seamless

continuity—much like individual frames in a movie film.

- In some species (like certain birds), this "frame rate" is higher, allowing them to perceive flickers and changes that humans entirely miss. In others (including humans), the rate is tuned for efficiency: not too fast to overwhelm, not too slow to miss crucial events.

2. The Mind's Stitching Function

- Your mind operates like an editor in a film studio—it takes these discrete field alignments and blends them into a coherent, flowing "cut" of reality.
- Small gaps between alignments are automatically bridged by memory, prediction, and expectation. This is why you can glance at a scene, blink, and still feel as though nothing was interrupted.
- This stitching process isn't perfect—it's where optical illusions, false memories, and inattentional blindness creep in.

3. Selective Attention as Alignment Filtering

- You do not align with *everything* in the field—only with what your current priorities, instincts, and ATT settings allow.
- Selective attention is essentially the dynamic gating of alignment bandwidth—you choose (consciously or unconsciously) which ripple handshakes to strengthen and which to ignore.
- This explains why you can "zone out" during a conversation and suddenly "snap back" when you hear your name—the alignment filter shifts, letting a new ripple pattern through.

4. Why Stability Is an Illusion

- The environment itself is in constant flux—temperature shifts, micro-vibrations, ripple disturbances—yet your perceptual system locks onto patterns it deems relevant, holding them steady in your mind's eye.
- The result is the *feeling* of a fixed, enduring world when, in reality, your awareness is sampling, selecting, and smoothing over countless micro-disruptions.

From a UED perspective, reality is not continuous—it is persistently re-negotiated. What you think of as "seeing the world" is actually your mind keeping up with a high-speed ripple conversation and presenting you with the *edited highlight reel.*

13.6 – Perception Gaps and Blind Zones

Mainstream science treats missed perceptions—like failing to notice something in plain sight—as errors or "limitations" in sensory processing. But from a UED perspective, they are not mistakes at all. They are the natural consequence of alignment rules. Your sensory field is not a passive, wide-open net; it is a targeted handshake mechanism. If a ripple pattern doesn't match your current alignment parameters, it may as well not exist for you—no matter how close, bright, or obvious it is.

1. Invisible Until Alignment Shifts

- Hidden objects, camouflaged creatures, and background movements all share the same core property: their ripple patterns fail to achieve the necessary coherence threshold with your field at that moment.
- The moment your alignment changes—perhaps due to a sound, a hint of motion, or a shift in attention—the same

object "suddenly appears," even though it never entered your space anew.

- This is why you can stare at a cluttered desk, fail to see your keys, and then "magically" spot them after you've already looked there twice. The keys weren't hiding; your alignment simply wasn't tuned to their ripple state.

2. Expectation as an Alignment Filter

- Expectation creates a ripple bias—you unconsciously tune your field to search for patterns that match what you believe is present.
- If you're looking for a "big red book," you may overlook the fact that the book you need is small and green, even if it's right in front of you.
- In extreme cases, expectation locks alignment so tightly that no contradictory pattern can break through—this explains why magicians, illusionists, and con artists exploit your focus to hide the obvious in plain sight.

3. Memory's Role in Shaping Perception

- Memory does not just store past experiences; it actively modulates present alignment by prioritizing patterns you've seen before.
- This explains "pattern blindness," where unfamiliar configurations are ignored in favor of familiar ones, even when the unfamiliar is more relevant.
- Your field, guided by memory, prefers to reaffirm the known rather than fully explore the new—an efficiency trick that sometimes costs you awareness.

4. Emotional State as a Ripple Amplifier or Suppressor

- Fear, excitement, calm, or stress alters the sensitivity of your alignment system.
- In fear, your field narrows—focusing tightly on threats, at the cost of peripheral awareness.
- In curiosity or play, your field broadens—allowing more varied patterns to align, making you more likely to spot the unexpected.

5. Inattentional Blindness, UED Style

- The famous "invisible gorilla" experiment—where participants fail to notice a person in a gorilla suit walking through a basketball game—perfectly illustrates alignment exclusion.
- In UED terms, the gorilla's ripple pattern never enters coherence with the observer's field because the active handshake is locked onto the ball-passing sequence.
- The phenomenon is not about missing photons—it's about never forming the ripple link in the first place.

From this perspective, "blind spots" are not sensory failures—they are alignment decisions, made consciously or unconsciously, in real time. And once you understand this, you realize that expanding perception is not about sharpening your eyes—it's about loosening the filters that dictate what your field will align with.

13.7 – Alignment Beyond the Senses

Perception is usually framed as a sensory act—light for the eyes, vibrations for the ears, molecules for the nose, and so on. But the UED model makes it clear that alignment is not bound to the five senses. Your field can achieve coherence with environmental ripple patterns without any sensory organ acting as a mediator. This is the realm where intuition, "gut feelings," and subtle impressions live—phenomena that mainstream science often treats as coincidence, bias, or superstition, but which UED frames as genuine field interactions.

1. Intuition as Direct Field Sampling

- Intuition is not a mystical sixth sense—it is alignment without translation.
- In sensory perception, a ripple pattern passes through a biological translator (eye, ear, skin) before reaching the mind's experiential space. In intuition, the field bypasses these intermediaries, forming a direct handshake with environmental states.
- This explains why intuitive impressions often feel instantaneous and complete—there is no gradual build-up of "evidence" in the sensory stream.

2. The "Gut Feeling" Mechanism

- The gut is a massive neural and electromagnetic hub, with its own semi-autonomous processing network.
- In UED terms, it acts as a low-latency alignment node, scanning ripple stability in your immediate environment.
- This is why you might feel an inexplicable tension upon entering a room where something is "off," even before anyone speaks or moves. The gut is reading the field, not waiting for visual or auditory confirmation.

3. Extrasensory Impressions and Ripple Overreach

- Sometimes alignment forms with ripple patterns outside your usual sensory range—emotional states of others, field disturbances caused by unseen movement, or subtle electromagnetic variations.

- You might "sense" someone behind you without hearing or seeing them because their field signature has already intersected with yours in the SVE structure.
- This is not about predicting the future—it's about reading ripple conditions before they manifest in the sensory channel you rely on most.

4. Sensing Atmosphere and Mood Without Cues

- Every group of people generates a shared field texture—a combined ripple pattern shaped by emotional states, interactions, and environmental factors.
- When you walk into a room and instantly know if it's tense, joyful, or awkward, you are not picking up on subtle visual clues alone—you are feeling the collective ripple alignment.
- This is why a room can feel "cold" or "warm" without any change in temperature—your field is responding to the emotional geometry, not the air molecules.

5. Why Mainstream Science Misses This

- Traditional science insists on sensory causation—if there's no measurable light, sound, or molecule to trigger a nerve, the perception is dismissed as imagination.
- UED reframes this: the field is primary, the senses are just one access point, and alignment can occur at any layer—physical, emotional, or cognitive—without requiring a sensory trigger.

6. Practical Implication

- If you accept alignment beyond the senses, you can train it—through mindfulness, environmental awareness, and emotional calibration.
- Skilled negotiators, athletes, and artists often rely on this form of perception, even if they don't consciously name it.
- The more sensitive your field becomes to ripple variation, the richer and more nuanced your perception becomes—sometimes revealing truths your eyes and ears can't yet confirm.

13.8 – Implications for Science, Technology, and Philosophy

When perception is reframed through UED as alignment rather than reception, the ripple effect extends far beyond biology. It changes how we design machines, how we study the brain, and how we think about the nature of reality itself.

1. Technology – Sensors That Detect Alignment, Not Just Physical Signals

- Current sensors are built to collect data streams—light intensities, sound pressures, temperature gradients. They assume perception is a matter of gathering enough raw input for later analysis.
- UED suggests a different design philosophy: create systems that detect field coherence states directly. Instead of measuring "how much" light or sound is present, they would recognize whether the ripple pattern matches a meaningful alignment signature.
- Such devices could work across mediums—detecting a "visual" pattern without visible light, or sensing "acoustic" structure without sound waves.
- This could open pathways to real-time situational awareness

in total darkness, underwater, or in deep space—environments where classical sensing is weak or impossible.

2. Neuroscience – Perception as a Participatory Event

- Mainstream neuroscience treats sensory processing as a pipeline: input enters through sensory organs, is processed in neural layers, and emerges as perception.
- UED's alignment model challenges this linear view. It suggests perception is co-activated—your neural field is not just processing incoming signals, but actively seeking and shaping alignment with the environment.
- This means the brain is not merely a passive receiver—it's a partner in reality formation.
- Research priorities would shift from analyzing stimulus–response chains to studying field readiness, alignment timing (ATT), and the neural conditions that optimize coherence events.

3. Philosophy – Reality as Co-Creation, Not Observation

- If perception is alignment, then the world you experience is never purely "out there." It is the outcome of a handshake between your field and the environment's field.
- This erases the traditional gap between observer and observed. Reality is not a fixed object waiting to be looked at—it is a participatory phenomenon that only takes form in the moment of alignment.
- Philosophically, this pulls perception away from the "mirror of reality" metaphor and toward the idea of mutual becoming—each act of perception changes both you and the world in subtle ways.

4. Wider Implications

- In ethics, this means perception always carries responsibility—because the reality you align with is partly your making.
- In education, training perception becomes as important as teaching knowledge—because sharper alignment means a richer and more accurate world.
- In art, it opens the possibility of designing works that operate not just on sensory input but on direct field resonance—art that can be "felt" even without being physically present.

Closing Reflection – "You Never Look at the World. You Meet It."

You have never truly *looked* at the world. Not in the way we are taught to imagine—standing apart, gazing at an external stage where life plays out. What you call "seeing" is not a transaction of photons and eyeballs. It is an encounter.

Perception is not the passive taking-in of reality. It is a meeting point—a handshake—between your field and the field of the world. In that instant, your internal structure and the external structure synchronize just enough for a shared reality to emerge. The colors, shapes, sounds, and textures you experience are not "out there" waiting for you; they *become* in that moment of alignment.

This means you don't live in a universe you simply observe. You live in a universe you **co-create** in real time, moment by moment, alignment by alignment. The stability you think you see is not fixed—it is the persistence of this handshake, renewed countless times a second.

And so, to "see" more of the world, you don't just open your eyes wider. You learn to open your alignment bandwidth—to tune yourself into a richer conversation with reality. The more you align, the more the world reveals—not because it has changed, but because *you* have stepped closer into the meeting.

In the end, life is not lived as a spectator sport. It is an ongoing, intimate dialogue between two realities—yours and the universe's—finding each other in the now.

Interlude 3: Chavan's Declaration

What if everything you've been taught about seeing the world is a clever illusion—an optical fairy tale designed to keep you thinking reality is "out there" while you stand here?

I, Sandeep Chavan, declare: You do not see through light. You do not see through eyes. You do not see through any "stream of photons" traveling obediently from object to organ.

You see because you **align**.

The miracle of vision is not in the catching of particles but in the synchronizing of patterns. Your eyes are not light traps—they are ripple translators, tuned to Structured Vacuum Energy. Every "image" is not carried to you like a postcard; it is co-created where your field and the world's field meet.

Color? Not wavelength. It is the *meaning* your mind assigns to a particular ripple signature. Reflection? Not a beam bouncing home, but a realignment handshake mediated by a surface. Refraction? Not speed changes, but ripple reorientation. Even shadows are not the absence of light—they are regions where alignment fails.

In this framework, perception is not a window—it's a **negotiation**. The world doesn't pour itself into you; you meet it halfway. Every sight, every hue, every shape is a temporary treaty between you and reality.

Why does this matter? Because once you see vision for what it is—a field dialogue—you stop being a passive receiver of the world and start being an active participant in it. You can learn to change what you perceive, not by altering the world, but by shifting your alignment.

This is not philosophy pretending to be science, nor science diluted into metaphor. This is **Universal Energy Dynamics**—a redefinition of optics, color, and perception from the ground up. If you accept this, you will never again believe that the universe simply "sends" you its truth.

You will know—truth is not sent. Truth is met.

And in every meeting, you are as responsible for what you see as the world is for showing it.

Part IV: The Cosmos and the Silent Field

Chapter 14: Light from the Stars

You Never Got It

Dismantling cosmic optics—no photons crossed the void

For as long as astronomy has told its stories, we've been sold a romance:

A photon is born in the heart of a star, fights its way to the surface, flings itself into the cold dark, and races for millions—sometimes billions—of years before, in a moment of cosmic intimacy, it strikes your eye.

It's poetic. It's cinematic. It's also wrong.

Under Universal Energy Dynamics, no photon has ever made that journey. You are not catching a relic from the past—you are participating in a *now* that happens to be coherent with a star's ripple signature. The starlight you "see" is not an ancient traveler—it is an instantaneous handshake between your local field and the star's field, mediated through Structured Vacuum Energy.

Once you understand this, the night sky changes forever. The stars are not beaming history at you; they are meeting you, here, now, in the same universe you occupy. The distance is not a journey—it is an alignment.

14.1 – The Romance of Starlight

For over a century, textbooks, documentaries, and planetarium narrators have painted the same picture:

Deep within the heart of a star, atoms collide with such ferocity that light is born. These newborn photons fight their way through dense plasma for thousands of years, finally escaping the star's surface. Then, they embark on a heroic voyage across the interstellar void—racing at 299,792 kilometers per second for millions, sometimes billions, of years—until, one quiet night, they end their epic journey in the back of your eye.

It's a beautiful story, and it resonates with us because it mirrors our own myths of endurance, travel, and destiny. We love the idea of touching something ancient, of receiving a literal piece of the past. The "light from Andromeda" becomes a kind of cosmic postcard, sent two million years ago and delivered just in time for you to notice it.

But there's a hidden cost to this romanticism. This narrative forces physics into uncomfortable corners—binding us to the speed-of-light limit, creating simultaneity paradoxes, and cementing the idea that what we see is always outdated, never truly "now." It makes the night sky a museum of fossils, not a living reality.

Worse, it turns "seeing" into passive reception of ancient messengers instead of what it truly is under Universal Energy Dynamics: an active, instantaneous alignment between fields.

The myth of photon travel may be poetic, but it locks us into a model that cannot fully explain why the universe looks and behaves the way it does in the present moment.

14.2 – The Photon Voyage Myth

The standard astrophysics story runs like clockwork in every classroom:

First, nuclear fusion in a star's core generates energy. This energy, we are told, takes the form of photons—tiny, discrete particles of light. But these photons don't leap straight into the cosmos. Instead, they endure an unimaginably slow "random walk" through the star's dense plasma, bouncing from one particle to another for thousands—sometimes millions—of years before they finally reach the surface.

Once free, these photons supposedly set off across the interstellar void, maintaining their identity like tiny, loyal couriers. They travel in straight lines for millions or billions of years, dodging cosmic dust, navigating gravitational fields, and never losing their structural integrity. At last, they reach Earth, slip through our atmosphere, and trigger the photoreceptors in your eyes or the CCD of a telescope.

It's a neat, linear chain of cause and effect:

Fusion → Photon Birth → Random Walk → Interstellar Voyage → Detection

But look closer, and the cracks show. Over such vast distances and timescales, space is not empty—it's a minefield of dust, gas, plasma, and radiation. The probability of a single photon surviving billions of years without being absorbed, scattered, or transformed is vanishingly small. Even in near-vacuum, quantum field fluctuations and cosmic interactions make the idea of a perfectly preserved photon absurd.

And yet, mainstream models require us to believe that *the exact photon* born in a star's core is the one arriving in your eye—pristine and unchanged. This is the same as claiming a soap bubble could survive a hurricane, a desert storm, and a volcano, all while maintaining perfect shape, and then drift into your hand intact.

The deeper flaw is conceptual: if photons are treated as persistent, intact travelers, they must be both perfectly isolated from the universe and yet perfectly interactive when they reach you. That is an impossible dual demand.

Under Universal Energy Dynamics, there is no photon ferrying itself across the abyss. What you experience as starlight is a present-moment field alignment event—your local field and the star's field synchronizing through the Structured Vacuum Energy

(SVE) continuum. No ancient package, no billion-year-old traveler—just an instantaneous handshake written into the field now.

The "photon voyage" is not an epic journey—it's a convenient metaphor that has overstayed its welcome.

14.3 – Stellar Ripple Signatures, Not Photon Streams

In the **UED framework**, stars are not cosmic light-bulbs firing streams of photons into the dark. They are *field generators*, continuously imprinting their unique ripple signatures into the Structured Vacuum Energy (SVE) matrix.

Every star's ripple signature is as distinctive as a fingerprint—shaped by its mass, composition, rotation, magnetic field, and internal dynamics. This pattern doesn't have to "travel" in the way mainstream physics imagines light beams darting through space. Instead, the ripple alignment spreads through the SVE in a way more akin to instantaneous field conditioning—a shift in the relational state of the whole continuum.

When you "see" a star tonight, you are not catching ancient photons that survived a cosmic marathon. You are locally decoding the current state of the ripple field at your position in space. The alignment between your perceptual field and the star's signature creates the visual experience we call starlight. It is an *event here and now*, not the delayed arrival of something launched long ago.

This understanding makes the concept of "light travel time" unnecessary. Mainstream astronomy insists that when you look at a star 1,000 light-years away, you see it as it was 1,000 years ago—because that's how long the photons supposedly took to get here. But in UED terms, there is no courier photon carrying a timestamp from the past. Instead, there is a present ripple state shaped by the star's ongoing interaction with the SVE, and your perception aligns with the current configuration of that field.

Think of it like a tuning fork: strike one fork in a connected medium, and another fork tuned to the same frequency will vibrate—not because a physical object traveled between them, but because the field they share was modulated. Stars "strike" the cosmic field; you resonate with that pattern in real time.

This also explains why the universe does not need to wait for unimaginably long "photon delivery schedules" to convey information about distant events. Instead of being at the mercy of the slowest possible speed of particles, perception happens when field conditions align, sidestepping the conceptual bottleneck of particle-based light travel altogether.

In this model, cosmic optics becomes less about beams crossing gulfs of space and more about reading the ripples in the shared fabric of reality. The star doesn't send you light—you and the star meet in the field.

14.4 – The Alignment Bridge

In the **UED view**, seeing a star is not the end point of a photon's million-year trek—it is the *moment* when your local field and the star's emission field achieve coherence. This moment of coherence is what we call the Alignment Bridge—a temporary, bidirectional handshake across the Structured Vacuum Energy (SVE) continuum.

The bridge is not a physical tunnel through which particles travel; it is a state of mutual ripple agreement. Your location and the star's position both exist inside the same universal field. When the ripple signatures match in frequency, phase, and tension resolution, they "lock in" to a coherent state. This lock-in is instantaneous—not because something moves infinitely fast, but because the underlying field is already continuous and connected.

Three critical factors govern the success of this bridge:

1. **Field Stability** – The SVE medium between you and the star must remain relatively undisturbed. Large-scale cosmic turbulence, plasma flows, or strong gravitational gradients can distort the ripple fidelity, weakening or severing the connection.
2. **Ripple Fidelity** – The integrity of the star's ripple signature must survive its propagation through the field. While the SVE is highly stable, it is not immune to interference. Other cosmic sources, overlapping ripple fields, and high-energy events can distort or mask the signal.
3. **Cosmic ATT (Alignment Threshold Time)** – Just as local perception requires an Action Threshold Time to lock onto an image, cosmic-scale alignments require their own ATT. This is the minimal temporal stability the field needs before your perception system registers the event as "seeing the star." If the alignment is too brief, the star will flicker, shimmer, or disappear.

This model also explains instantaneous disappearance. In the photon-travel narrative, a distant star that suddenly ceases to exist should continue shining in our sky for however many light-years away it is—its "in-flight photons" supposedly still on their way. But under UED, break the alignment bridge and the star's visibility ends immediately.

Phenomena like gravitational lensing shifts fit naturally here. When a massive object moves between you and the star, it alters the field geometry. The star doesn't get "bent around" in a photon arc—it simply loses direct alignment with you, and your visual lock breaks until the geometry stabilizes again.

This also accounts for certain mysterious astronomical events—brief, unexplainable disappearances of distant stars and galaxies recorded by telescopes. In UED terms, these are not cosmic catastrophes or "failed observations" but alignment disruptions—a ripple handshake lost in the noise of the field.

When you look up at the night sky, you're not watching ancient messengers arriving from across the void—you're standing on the edge of countless alignment bridges, each one holding just long enough for you to meet a star in the eternal present.

14.5 – The Illusion of 'Looking Back in Time'

Astronomy loves the poetic claim that "When you look at a distant galaxy, you're seeing it as it was billions of years ago." It's a powerful narrative—one that links human eyes to deep cosmic history. The logic is rooted in the photon-journey model: if light takes billions of years to cross the void, then what enters your telescope tonight must have left its source in the remote past. You are, so the story goes, not looking into space—you're looking back through time itself.

Under UED, this is a misinterpretation of causality. You are not receiving an *old parcel* of photons preserved for eons like an undelivered letter. You are sampling a present field state—a ripple configuration in Structured Vacuum Energy (SVE) that still carries the source's imprint. That imprint is not a "frozen particle" from the past; it's an active signature sustained by ongoing coherence between your location and the source's ripple pattern.

The difference is subtle but profound:

- In the photon model, the "past" is a physical delivery arriving late.
- In UED, the "past" is a meaning we infer from the ripple's pattern—its frequency content, decay rates, and interaction history with the field.

In other words, temporal inference is not the same as actual past observation. You deduce that a star is old, or that a supernova happened millions of years "ago," not because you are literally seeing that moment, but because the pattern you align with today contains structural features consistent with such an event in its field evolution. It's like reading tree rings: the rings tell you about past seasons, but you are not traveling to those seasons—you are reading their record in the present.

This distinction resolves several contradictions in the look-back narrative:

- **Instant disappearance problem** – If a galaxy truly vanished billions of years "ago," the photon model says we'd keep seeing it until the last photons arrive. In UED, break the alignment and it vanishes immediately—no "light still in transit" required.
- **Interference with supposed ancient photons** – Cosmic dust, plasma fields, and gravitational distortions would scatter or erase ancient photons long before they reached us. In UED, the signal is maintained by present-time coherence, not impossible long-term survival.
- **Simultaneity paradox** – The photon model demands strict sequencing of events across distances. UED sidesteps this entirely—what you see is always "now," even if the pattern encodes a long causal history.

So when you gaze at Andromeda and think, *"I am seeing it 2.5 million years ago,"* you're engaging in a poetic metaphor, not a literal act. You are meeting Andromeda now, through a ripple alignment whose structure whispers stories of its long cosmic journey—not because those stories traveled here, but because they are woven into the living present of the field.

14.6 – Why Stars Twinkle

In the mainstream picture, the twinkling of stars—astronomers call it *stellar scintillation*—is explained as an atmospheric problem. The story goes like this: photons from a star enter Earth's atmosphere, and as they pass through layers of air at different temperatures and densities, the air's refractive index shifts unpredictably. This bends and scatters the photons' paths, causing the apparent position and brightness of the star to fluctuate from moment to moment.

While this account works within the photon-travel model, it also inherits that model's weaknesses—chief among them, the assumption that light is a stream of particles (or waves) enduring a gauntlet of atmospheric obstacles before reaching your eye.

Under UED, twinkling is not the *misdirection of incoming particles*. It is the real-time modulation of ripple alignment coherence between your local field and the star's field, mediated through the structured vacuum energy (SVE) network.

Here's what's actually happening:

- A star's ripple signature is coherent enough for your field to align with it, producing a stable perception.
- The atmosphere acts as a dynamic ripple filter, not a photon deflector. Thermal gradients, wind shear, and density variations all shift how local SVE ripples couple with the star's distant ripple pattern.
- When the atmospheric state changes rapidly—millisecond-scale in many cases—it perturbs the stability of the alignment handshake. Your perception registers this as tiny, rapid changes in intensity and position: the star "twinkles."

This model naturally explains phenomena the photon view struggles with:

- **Wavelength-specific twinkle** – Blue and red light from the same star often twinkle differently because each wavelength corresponds to a slightly different ripple mode. The atmosphere's state may momentarily favor alignment with one mode over another, altering perceived color or brightness independently.
- **Altitude effects** – Stars near the horizon twinkle more because their ripple alignment passes through more atmospheric volume, increasing the number of unstable ripple interactions before reaching coherence. Near the zenith, alignment is more direct and stable.
- **Planet vs. star behavior** – Planets rarely twinkle to the naked eye because they are close enough for your field alignment to average across a broader angular ripple spread. Stars, by contrast, are point sources—your field handshake is much narrower and thus more easily perturbed.

Seen this way, twinkling isn't a light-path accident—it's a field relationship under stress, like trying to shake hands with someone while a crowd jostles between you. The handshake never truly breaks, but it wavers, tightens, loosens, and occasionally slips, producing the charming sparkle that poets have written about for centuries.

14.7 – No Photon Loss, Only Alignment Loss

In the conventional model, when a star appears dimmer near the horizon, the explanation is straightforward: photons have to pass through a thicker layer of atmosphere, where many are absorbed or scattered away from the line of sight. The longer the path, the greater the loss, so the star looks weaker.

While this fits within the particle-travel story, it also assumes that light is a finite commodity moving through space like cargo on a road, with "units" being lost along the way.

Under UED, nothing of the sort happens—because nothing is "arriving" in the first place. Your perception of the star is not the cumulative count of photons that survive the journey. It is the instantaneous coherence event between your local ripple field and the star's ripple signature in the structured vacuum energy (SVE) matrix.

Here's what changes near the horizon:

- **Alignment path instability** – The handshake between your field and the star's field must pass through more atmospheric volume. Each layer of turbulence, density gradient, and temperature variation shifts the local SVE ripple conditions.
- **Ripple distortion accumulation** – These small disruptions don't block photons; they erode the fidelity of the ripple pattern. Think of it like a radio signal passing through increasingly noisy interference—the message isn't stopped, but it becomes harder to decode.
- **Threshold effects** – If the pattern fidelity drops below your perceptual ATT (Alignment Threshold Time) requirements, the star's presence fades. This is not dimming by energy loss—it's dimming by handshake breakdown.

This view transforms our understanding of deep-space observation:

- **Fading is not about cosmic depletion** – Even light from extremely distant galaxies doesn't "wear out" or vanish with distance. What limits visibility is the stability of long-range field alignment, not the endurance of individual photons.

- **Clarity can be restored without collecting more light** – In principle, future telescopes could boost visibility by stabilizing field alignment rather than increasing aperture size. The "light bucket" concept could be replaced with ripple coherence stabilizers.
- **No energy exhaustion across space** – In UED, there is no cosmic energy leak due to transit; the field is everywhere. What changes is our ability to maintain an undistorted relationship with a distant ripple pattern.

From this perspective, stars don't fade because they are "running out" of visible light before they reach you. They fade because the conversation between your field and theirs gets too garbled to continue—until alignment is either restored or the star rises higher, clearing the noisy horizon path.

14.8 – Cosmic Optics Without Travel Time

If light is not a courier that departs from a distant star and spends millions—or billions—of years crossing the void, then much of modern astronomy's optical framework has to be rebuilt from the ground up. In UED, observation is an alignment event between your local ripple field and the source's ripple field in Structured Vacuum Energy (SVE). This shift immediately dismantles the time-travel narrative that underpins redshift, cosmic distances, and the age of the universe.

1. Rethinking Redshift – From Doppler to Drift

In the conventional view, redshift is interpreted as a Doppler effect—if a galaxy's light is stretched toward the red end of the spectrum, it must be receding from us at high speed. The further the shift, the faster it is moving, and by extrapolation, the older the universe must be.

In the UED view, there is no physical beam in motion to be stretched. The "red" we detect is not a stretched photon—it is a current ripple state of the field, shaped by interactions between the source's signature and intervening cosmic structures.

- **Ripple State Drift** – Over vast spatial separations, local and cosmic-scale field conditions evolve, shifting the stable alignment frequency between observer and source. This manifests as a spectral shift without implying any velocity of separation.
- **No Recession Required** – A redshift can arise purely from the evolution of the field's structural tension—not from galaxies speeding away into expanding emptiness.

2. Implications for Cosmic Distance Measurement

Modern astronomy uses redshift as a ruler: more redshift = greater distance. This assumes the only cause is recessional velocity over time. But if redshift is partly or entirely field drift, then our "yardstick" is not fixed—it's mood-sensitive.

- **Distance estimates could be inflated** – Some "most distant" objects may not be physically that far, only strongly affected by ripple drift conditions.
- **Cosmic mapping becomes relational** – Instead of measuring how far light traveled, we would chart how stable field alignments vary across the universe.

3. Expansion Theory Under Pressure

If redshift is not a velocity indicator, then the Big Bang expansion model loses its primary observational anchor. The notion of an inflating spacetime fabric could be replaced with a dynamic field equilibrium model, where large-scale ripple patterns shift over time without requiring galaxies to accelerate apart.

- This removes the need for exotic "dark energy" to explain accelerating expansion.
- The cosmic web could be a standing field configuration, not a ballooning geometry.

4. The Age of the Universe—A Question, Not an Answer

In mainstream cosmology, light travel time is the clock: if we see a galaxy at 13 billion light-years, it must be 13 billion years old. But under UED, you are never seeing an ancient object—you are encountering its present field state.

- The universe's "age" becomes unpinned from the light travel metric.
- Cosmology shifts from archaeology to real-time field dynamics.

Closing Thought: Without photon travel time, cosmic optics stops being a courier service and becomes a real-time conversation between here and there. Redshift becomes an accent in the cosmic dialogue, not a speedometer reading. Distance becomes a measure of alignment stability, not elapsed travel. And the age of the universe? That might be the wrong question entirely—because in the field, everything is now.

14.9 – Technological and Scientific Shifts

If UED's interpretation of starlight is correct—where no photons cross the void and all cosmic "seeing" is an instantaneous alignment event—then both our tools and our scientific strategies will need a fundamental redesign. The implications are not just incremental; they are paradigm-shattering.

1. Instruments Designed for Alignment Optimization, Not Photon Capture

Traditional telescopes are built on the assumption that light is a stream of particles or waves arriving from a distant source. The bigger the mirror or lens, the more "light" you collect.

- Under UED, collection is irrelevant—you're not scooping up photons, you're tuning to a ripple state.
- Future instruments would resemble cosmic resonance tuners rather than optical buckets—configurations of field-sensitive detectors designed to maximize coherence lock with a target's ripple signature.
- The metric of success would shift from "aperture size" to alignment precision and stability across SVE conditions.

2. Real-Time Mapping of Stellar Fields Without Massive Optics

If no light travels, then a telescope's large mirrors are not gateways to distant photons—they are unnecessary for basic field access.

- Compact, field-alignment devices could detect and decode stellar signatures in real time, potentially even from platforms much smaller than current observatories.
- Distributed detector arrays could map the sky with unprecedented speed and resolution by synchronizing their alignment states—essentially building a planet-wide coherent eye without a single massive instrument.
- This approach could democratize high-precision astronomy, making "deep space observation" possible from small-scale labs, even handheld devices.

3. Rethinking Interstellar Communication

Current communication models assume a signal travel time—radio waves or laser beams that cross the void at finite speeds.

- If alignment events can be established between two distant points in SVE, communication becomes an act of instantaneous field engagement.
- No carrier particles need to traverse the gap; instead, information is embedded into a ripple state that both ends share in real time once coherence is established.
- This could render light-years of delay obsolete, collapsing the gap between interstellar civilizations into a now-based exchange.
- Security and privacy would also transform—without a traveling signal to intercept, eavesdropping becomes exponentially harder.

4. New Scientific Paradigms

The shift from photon traffic to field alignment would ripple into every branch of astrophysics:

- Spectroscopy would measure present alignment drift rather than reconstructing ancient light.
- Cosmology would focus on current field topology, not historical travel-time maps.
- Planet hunting could be done by detecting subtle field distortions from orbiting bodies, bypassing transit photometry and Doppler wobble entirely.

Closing Note: In this new framework, the telescope becomes less of a "cosmic bucket" and more of a cosmic handshake device. Astronomy stops being about catching ancient messengers and becomes the art of meeting the stars where they are—right now. The day we fully embrace this will be the day our observatories shrink, our reach expands, and the word *distance* takes on an entirely new meaning.

Closing Reflection – "The Stars Are Not Sending You Their Past"

We have been told a romantic story—that when we gaze at the night sky, we are looking back in time, catching the final photons of a journey that began millions or billions of years ago. It's a poetic picture: ancient light, finally arriving, slipping into our eyes after an odyssey across the void. But under UED, this picture dissolves.

You are not a distant light collector. You are not a passive recipient of cosmic deliveries. You are an active participant in a present-tense alignment, a local tuning into a shared ripple field that exists here and now. The "starlight" you experience is not a courier's parcel from a forgotten past—it is the immediate consequence of your field and the star's field finding temporary coherence in the vast, structured vacuum.

Every glimmer you perceive is a meeting, not a memory—a handshake across the cosmos, forged in the instant of awareness. The past does not travel to you; it is inferred from patterns in the present. The stars are not sending you their history. They are standing with you, in the same universal now, speaking in a language that only alignment can translate.

Once you see this, the sky changes. It stops being a museum of fossil light and becomes a living network—one where every star, every galaxy, is connected to you in real time. The wonder remains, but it shifts from the romance of travel to the intimacy of presence. You are not watching from afar. You are already there.

Chapter 15: Redshift Rewritten

Ripple Delay and Misalignment, Not Velocity or Expansion

For over a century, redshift has been the astronomer's speedometer—a silent signal that galaxies are racing away, dragging space itself into an ever-expanding void. Entire cosmologies, from the Big Bang to dark energy, rest on the simple assumption that "longer wavelengths mean faster recession."

But what if that assumption is wrong?

What if the universe is not stretching, and nothing is fleeing? What if the "cosmic Doppler" you've been told about is not a measure of motion at all—just a misreading of how fields talk to each other?

Under Universal Energy Dynamics (UED), redshift is no tale of runaway galaxies. It is the fingerprint of ripple delay—a small, precise misalignment between your field and the field of the object you observe. A wavelength can stretch without a single atom moving an inch. A cosmic spectrum can drift without any star or galaxy having traveled toward or away from you.

Once you strip away the romance of an expanding universe, redshift stops being a story about the past—and becomes a tool for reading the present condition of the cosmic field itself. And that changes everything.

15.1 – The Standard Redshift Story

In mainstream astrophysics, redshift is treated as the universe's universal speedometer—a direct measure of how fast something is moving away from us or how much the space between us has expanded. The foundation is the Doppler effect, familiar from the change in pitch when a siren passes you: if the source moves away, waves stretch; if it moves closer, they compress. Apply this to light, and stretching its wavelength pushes it toward the red end of the spectrum—hence, "redshift."

The standard narrative categorizes redshift into three main types:

1. **Cosmological redshift** – The crown jewel of Big Bang cosmology. This assumes that the fabric of space itself is stretching, carrying galaxies along for the ride. Light from distant galaxies is stretched *because the space it travels through is expanding*. The further away the source, the greater the redshift—supposedly proving the universe is expanding uniformly.
2. **Doppler redshift** – A straightforward velocity interpretation. If a star or galaxy moves away from you, the light's wavelength elongates, shifting toward red. This is common in local motions, like stars orbiting within galaxies or galaxies interacting within clusters.
3. **Gravitational redshift** – Predicted by Einstein's general relativity. Here, light loses energy climbing out of a gravitational well. The loss doesn't slow it down—it can't—but it supposedly stretches its wavelength. This effect is small but measurable near extremely massive bodies like neutron stars or black holes.

The unifying assumption behind all three is photon travel. In this view, photons leave a source, carry their energy (and frequency) intact across vast distances, and arrive at your detector altered only by motion, expansion, or gravitational influence encountered along the way.

This makes redshift a history book—every photon is treated as a courier that preserves a record of the moment it left its source, allowing astronomers to "look back in time." Cosmological redshift, in particular, underpins the Big Bang model, dark energy hypotheses, and the estimated age of the universe.

But this model has cracks. It depends on untestable assumptions about the nature and rate of cosmic expansion. It insists that the *only* acceptable explanation for systematic redshift is motion or stretching space—rejecting older ideas like "tired light" (where photons lose energy over distance) on the grounds that they can't explain certain observational patterns, while ignoring inconsistencies of its own.

The deeper issue is that all three redshift types require the belief that photons survive an uninterrupted journey across billions of light-years, through dust, gas, gravitational fields, and cosmic plasma, without scattering into noise. Yet the same physics tells us such a voyage would be wildly improbable without massive distortion or loss.

UED breaks from this entirely—not by adding another "tired light" variant, but by removing the travel premise altogether. In our framework, redshift is not a record of motion—it's a symptom of field misalignment in the present. The color shift is real, but the story behind it changes completely.

15.2 – The Core Contradictions

The standard redshift framework—where wavelength stretching is tied to velocity, expansion, or gravitational escape—sounds mathematically neat, but the sky refuses to behave that neatly. The moment you look beyond textbook diagrams, you find observations that don't just bend the model—they break it.

One of the most glaring problems comes from anomalous redshifts. Astronomers have repeatedly observed cases where two celestial objects appear physically connected—sharing material bridges or gravitational interactions—yet display wildly different redshift values. A classic example is quasars observed near galaxies: the quasar might show an extreme redshift, suggesting it is billions of light-years farther away, while the galaxy's redshift implies a much closer location. If the two are truly interacting, their distances can't be so mismatched. The standard model's escape route here is to call it coincidence or line-of-sight overlap—but the frequency and clarity of these cases strain that excuse.

Then there's the fact that redshift is not always proportional to inferred velocity or distance. According to the Hubble Law, the relationship should be clean: higher redshift equals faster recession and greater distance. But numerous surveys show objects at similar redshifts with vastly different inferred distances, and vice versa. The further out we look, the more scattered the redshift–distance relationship becomes, suggesting that it's not the universal ruler it's claimed to be.

The expansion paradox is another thorn. If galaxies at high redshift were truly receding at the speeds required by the expansion model, the early universe should have been far less structured—chaotic, clumpy, and without the intricate spiral patterns we see today. Yet we find high-redshift galaxies that are mature, aligned, and well-organized, as though they had billions of years to settle into order—directly contradicting the supposed timeline. This forces cosmologists to invent patchwork explanations like "early rapid evolution" or "galaxy assembly bias," none of which emerge naturally from the original model.

Even gravitational lensing—the bending of light by massive objects—runs into trouble when redshift is treated purely as a velocity shift. In some lensed systems, the redshift values of the source and the lens don't line up with their supposed distances, forcing convoluted "foreground–background" rearrangements that often feel more like model-saving than truth-finding.

In short, redshift in the conventional framework behaves like a ruler that measures accurately only when you already know the answer you want. Once you strip away the narrative scaffolding, the inconsistencies point to a deeper truth: redshift might not be telling us about travel or speed at all.

In UED terms, these contradictions vanish—not because we ignore them, but because we stop assuming that light is a physical courier crossing the void. Instead, we treat redshift as a present-moment field effect—a change in ripple alignment state—which means mismatches, non-proportionality, and structural coherence are not anomalies but expected consequences.

15.3 – UED's Ripple Redshift Principle

In the Universal Energy Dynamics (UED) framework, redshift is stripped of its baggage as a speedometer for the cosmos. Instead, it becomes something far simpler—and far more consistent with observation: a measure of alignment drift between fields.

The starting point is this: you are not receiving a "packet" of light from a moving source. You are engaging in a field-to-field alignment event—a moment where your local Structured Vacuum Energy (SVE) field and the star's or galaxy's field achieve coherence. What we call "redshift" is simply what happens when that coherence is slightly out of phase compared to the source's native signature.

This phase difference arises from ripple delay—the stretching of interaction patterns when fields do not match perfectly in their oscillatory relationship. Think of two musical notes that should be in harmony but are slightly detuned; the interference pattern between them creates a slower, longer wave envelope. In optical terms, this manifests as a shift toward the red without requiring any change in the physical frequency of a traveling photon—because there is no traveling photon.

What causes this misalignment? Not the recession of a galaxy at unimaginable speed, but field disturbances and medium changes along the alignment bridge. Intervening matter, gravitational wells, plasma regions, and even large-scale cosmic field turbulence can slightly alter the phase relationship between the source's ripple signature and the observer's local field. These are not "obstacles" to a traveling beam—they are structural modifications to the shared medium of SVE, adjusting the conditions under which alignment is achieved.

This is why ripple redshift can happen with no relative movement at all between the source and the observer. Two perfectly stationary bodies, separated by vast space, can still experience a measurable redshift if the intervening field conditions change over time. A shift in cosmic ATT (Alignment Threshold Time), a disturbance from a passing gravitational wave, or a density fluctuation in intergalactic plasma can all alter the moment-to-moment phase matching, stretching the observed wavelength.

By decoupling redshift from velocity, UED immediately solves multiple cosmological puzzles:

- Quasar–galaxy anomalies become natural—different alignment paths mean different ripple delays.
- Non-proportionality with distance is expected, because redshift measures field condition history, not space covered.
- Apparent "look-back time" collapses into present-moment alignment sampling, freeing us from the paradox of mature

galaxies existing "too early."

Under Ripple Redshift, the universe stops racing away from us in an endless expansion—it becomes a dynamic, stable field system, where redshift is not the ticking of a cosmic speedometer but the weather report of inter-field conditions.

15.4 – The Role of Cosmic ATT

In UED, Alignment Threshold Time (ATT) is not just a local sensory concept—it is a cosmic-scale regulator of what we perceive as frequency, wavelength, and color in starlight and galactic light. ATT defines the minimum duration required for two fields to achieve a coherent exchange—whether that's between your eye and a nearby candle flame or between a radio telescope and a quasar five billion light-years away.

When ATT shifts, the *effective phase matching* between your local SVE field and a distant source's ripple signature also shifts. Even a microscopic delay—fractions of an oscillatory cycle—can cause the *perceived* wavelength to stretch. This is not because energy is "bleeding away" in transit, nor because the signal is "slowing down," but because the alignment itself is happening fractionally later in the cycle than it would in perfect phase lock.

Imagine tapping two drums in perfect rhythm. If one drummer begins hitting just slightly late each time—not enough to create an echo, but enough to soften the beat—the combined pattern you hear changes. The tempo feels slower even though neither drummer has actually slowed their internal pace. In optical terms, this manifests as a redshift: the wave interactions appear longer, though the intrinsic source pattern remains unchanged.

Cosmic ATT is not static. It can vary due to:

- Large-scale gravitational influences bending or distorting

field relationships.
- Density fluctuations in cosmic plasma and interstellar mediums altering ripple stability.
- Long-period field oscillations that subtly modulate alignment readiness across vast distances.

This variation in ATT explains phenomena that have puzzled astronomers for decades:

- **Redshift clustering** – Certain redshift values appear more often than others in galaxy surveys. In UED terms, these are not cosmic "speed zones" but preferred ATT states where large regions of the cosmic field enter similar alignment readiness, producing repeating bands of ripple delay.
- **Quantized redshift** – The controversial observation that some galaxy redshifts occur in discrete steps rather than continuous gradients. Mainstream physics struggles to reconcile this without invoking measurement errors or obscure statistical coincidences. UED instead predicts that when cosmic ATT fluctuates in harmonics, phase offsets occur in discrete intervals, naturally creating apparent redshift quantization.
- **Apparent large-scale structure correlations** – When galaxies over huge distances appear more synchronized in redshift than expansion models allow, it's because their shared field environment is modulating ATT in unison—not because they are all moving at the same velocity away from us.

In this view, the "speed" of the universe becomes irrelevant. What matters is how ready the cosmic field is to align with you at any given moment. The moment of readiness *is* the moment of perception, and any delay or phase offset in that readiness is immediately translated into a color shift.

15.5 – Gravitational & Environmental Influences

Mainstream physics frames gravitational redshift as a photon energy loss—the idea that light must "climb out" of a gravity well, losing frequency along the way. In UED, no such uphill journey occurs, because no photons are traveling. Instead, what changes is how the source field and the observer's field align when embedded in different gravitational and environmental conditions.

Field Curvature as Ripple Alignment Shift

Massive bodies—stars, planets, black holes—don't "bend spacetime" in the geometric sense Einstein described; they reshape the local Structured Vacuum Energy (SVE) field. This reshaping alters ripple geometry and timing, meaning the wave interaction between you and the source must resolve over a slightly altered alignment path.

- Near a massive body, field curvature modifies the phase reference, shifting the perceived oscillation rate when the handshake with your local field is established.
- This shift can stretch the perceived wavelength (redshift) when the source is deeper in the curvature than you, or compress it (blueshift) if you are deeper and the source is higher.

Environmental Drift from Medium Density

Even in deep space, the medium is never truly empty. Interstellar plasma, cosmic dust, and varying SVE density gradients impose small but cumulative changes in ripple coherence.

- These act as slow filters—not absorbing photons, but modulating when alignment occurs along a ripple cycle.
- Over millions of light-years, even tiny misalignments per region stack into significant perceived redshifts or blueshifts, depending on whether the medium lags or advances the alignment readiness.

Gravitational Redshift Reinterpreted

Instead of imagining light "losing energy" while escaping a gravity well, UED sees gravitational redshift as the difference in field reference states between the source location and the observer location.

- Inside a deeper gravitational gradient, SVE ripple rates are offset relative to a shallower region.
- When the handshake occurs across these gradients, the alignment interval (ATT) is different, making the observed frequency appear shifted.
- The wave source's intrinsic frequency never changes—it is the cross-field phase agreement that has shifted.

Why This Matters

This view eliminates the photon survival problem entirely. Nothing must physically travel intact across millions of light-years; instead, coherence is re-established locally. It also explains why gravitational redshift measurements match predictions without invoking an "energy drain" process—and why shifts can occur even in situations where no relative motion exists between source and observer.

15.6 – Redshift Without Motion: Stationary but Shifted

The mainstream model makes velocity the centerpiece of redshift interpretation—if something is redshifted, it must be moving away. Under UED, this assumption collapses. Redshift can occur without any physical movement of the source or observer because it is not a velocity effect; it is a field alignment effect.

The Stillness Principle

In UED, the universe is fundamentally still. Objects are not flinging themselves away at cosmic speeds; instead, ripple fields are continuously reconfiguring.

- A redshift can emerge purely from misalignment drift between the source field and the observer's field.
- This drift might be caused by changes in the intervening medium, local SVE density variations, or gravitational gradients—even when both source and observer are fixed in position relative to one another.

Laboratory Analogues

Experiments already hint at this possibility:

- **Temperature-induced spectral drift**: A stationary lamp's emission spectrum shifts slightly as the surrounding medium heats or cools—not because the lamp moves, but because alignment conditions in the medium change.
- **Index-of-refraction variations**: In controlled optical setups, altering the refractive index of a medium between source and detector causes measurable wavelength shifts without any source motion.

In UED terms, both examples show how altered ripple coherence timing can stretch or compress the observed frequency without invoking Doppler mechanics.

Cosmic Examples

Several astronomical observations fit the "stationary but shifted" profile:

- **Quasars with anomalous redshifts**: Some quasars appear gravitationally bound to nearby galaxies yet have vastly different redshifts, which in the Doppler model would require absurd relative velocities.
- **Galactic cluster redshift anomalies**: Neighboring galaxies sometimes exhibit redshift differences larger than their gravitational binding should allow, suggesting alignment drift rather than physical separation speed.
- **Gravitational lens redshift mismatches**: Light from the same source can be bent by a lensing galaxy into multiple paths that show slightly different redshifts—impossible if the shift were purely from source motion.

The Ripple Drift Mechanism

Under UED, stationary redshift emerges from delayed ripple phase matching:

1. The source's ripple signature propagates as a pattern in the SVE, not as traveling photons.
2. If the intervening field undergoes structural change—density fluctuation, turbulence, or gravitational curvature—the timing of phase-lock between source and observer fields shifts.
3. This altered Alignment Threshold Time (ATT) modifies the observed oscillation rate, producing a measurable

redshift or blueshift without motion.

Why This Redefines Cosmology

If redshift does not require movement, then:

- Distance scaling laws based on Doppler velocity collapse.
- The Hubble constant no longer directly measures "expansion speed."
- Cosmic age estimates tied to redshift ratios become questionable.

This reframing also removes the need for speculative concepts like dark energy—much of what we call "accelerating expansion" may simply be alignment drift accumulation over large-scale structures.

15.7 – Rethinking the Hubble Law

For over a century, the Hubble Law has been the cornerstone of the expanding-universe narrative: the farther away a galaxy is, the greater its redshift, implying it's receding faster. In the mainstream view, this linear relationship is interpreted as *direct evidence* of cosmic expansion, with Hubble's constant serving as the yardstick for the universe's growth rate.

UED turns this interpretation inside out—not by denying Hubble's data, but by reframing its meaning entirely.

Hubble's Observation as a Statistical Field Pattern

Hubble's original finding—that redshift increases with distance—does not *require* velocity. It simply records a correlation between separation and spectral shift. Under UED, this correlation is the natural result of statistical ripple misalignment across large-scale field separations.

- As the alignment distance between two points in the cosmic field increases, the probability of perfect phase matching decreases.
- This reduced coherence manifests as a systematic shift toward lower observed frequencies—what we currently label as "redshift."
- The relationship appears linear *on average* because, for moderate ranges, the cumulative misalignment tends to increase in roughly uniform increments.

In other words: Hubble's law may be a statistical fingerprint of ripple drift—not a speedometer for galaxies.

Ripple Misalignment Scaling with Observational Separation

In UED, field alignment operates like a handshake across the structured vacuum. The farther apart two points are:

- The more intervening field structures (galaxies, voids, plasma regions) the alignment must traverse.
- Each of these structures introduces a small phase drift—not because of motion, but due to *local variations in field density and structure.*
- Over vast distances, these small shifts add up to a measurable spectral displacement, mimicking the appearance of a steady recession velocity.

Why Hubble's "Constant" Might Not Be Constant at All

If redshift is an alignment effect, Hubble's constant is not a fundamental property of the universe—it's an *emergent ratio* from a snapshot of our current cosmic conditions.

- In regions with different field drift distributions (e.g., cosmic voids vs. galaxy-rich areas), the slope of the redshift-distance relationship could change.

- Over cosmic time, as large-scale structures evolve, the observed value of H_0 should vary—exactly what recent surveys have found, leading to the so-called *Hubble tension.*
- This variation is no longer a paradox under UED—it is a predictable outcome of a dynamic, ripple-structured universe.

In this light, Hubble's Law is not the voice of a universe stretching itself thinner—it's the statistical hum of a cosmic field that is never perfectly aligned over great distances. The "constant" is simply today's average level of large-scale ripple drift, not a fixed expansion rate.

15.8 – Testing the Ripple Redshift Model

A theory is only as strong as the experiments that can challenge or confirm it. If UED's Ripple Redshift Principle is correct, it should produce testable deviations from the velocity-based model of redshift. These deviations wouldn't be subtle footnotes—they'd be clear, measurable patterns that mainstream cosmology currently struggles to explain.

Predictable Differences in Redshift Under UED

Non-linearity Over Extreme Distances

- In a purely expansion-based model, redshift should scale in a predictable way with distance (once local motions are accounted for).
- Under UED, ripple misalignment compounds *non-uniformly* at great separations, producing curvature in the redshift-distance graph for extremely distant objects.
- This non-linearity is already hinted at in deep-field surveys where very high-redshift galaxies appear brighter, larger, or more structured than the expansion model allows.

Redshift Shifts Without Velocity Change (Lab-Scale)

- If ripple drift can occur without motion, then we should be able to replicate redshift *in a stationary system*.
- In a controlled lab, by altering the local field alignment between two stationary points (using ripple-interference methods), we should measure a shift in observed frequency without any Doppler movement.
- This would directly falsify the assumption that redshift always means recession.

Anomalies in Lensed Objects' Redshifts Explained by Field Distortion

- Gravitational lensing often shows multiple images of the same source with slightly different redshifts, a paradox for the standard photon-travel model.
- Under UED, each lensed path passes through a slightly different field density profile, altering ripple alignment and producing distinct spectral shifts *without requiring multiple sources or variable velocities.*

Proposed Experimental Setups

Earth-Based Ripple Alignment Chambers

- Construct controlled environments where Structured Vacuum Energy (SVE) density and ripple conditions can be modulated.
- Use laser systems or coherent EM sources placed in fixed positions, then vary the intervening ripple state to induce measurable frequency shifts.
- This would directly demonstrate alignment-induced spectral changes without motion.

SVE Phase Modulation Detection

- Develop sensors capable of detecting minute phase drifts in a field—essentially measuring the *alignment quality* rather than just photon flux.
- These could be deployed in astronomical observatories to compare redshift values under different field alignment states (e.g., during geomagnetic storms vs. calm periods).

Gravitational Lens Field Profiling

- Use known strong-lensing systems to map redshift variations across multiple images of the same object.
- Model the lens not as a bending surface for photons, but as a ripple-density modulator, predicting the exact spectral differences between each image.

If these experiments confirm even one of the above predictions—especially stationary redshift shifts in the lab—the Doppler-expansion interpretation of the universe would no longer be the default. Instead, we would enter a new era of cosmic field spectroscopy, where distance, motion, and even the age of the universe are re-evaluated through the lens of alignment physics.

15.9 – Implications for Cosmology

If UED's interpretation of redshift is correct—if it's primarily a ripple alignment drift rather than a velocity indicator—then much of modern cosmology's scaffolding comes crashing down in an instant. This isn't a minor recalibration; it's a wholesale rewrite of the cosmic story.

The Collapse of the Velocity–Expansion Paradigm

- The entire Big Bang model hinges on interpreting redshift as evidence that galaxies are receding at measurable velocities. Remove velocity from the equation, and the central claim—that the universe began in a singular explosive expansion—loses its primary observational pillar.
- Age estimates based on Hubble's Law also evaporate. If redshift is not proportional to recession speed, then "looking back" billions of years to a hypothetical origin is no longer a valid inference. The universe could be vastly older—or even timeless in its underlying structure.

Alternative Cosmic Models Under UED

- **Steady-State Ripple Universe** – The large-scale structure could be stable in its fundamental framework, with observed changes being local reconfigurations of field alignment rather than expansion or contraction.
- **Cyclic Alignment Patterns** – Apparent large-scale shifts (such as redshift drift) might follow rhythmic, long-term cycles in the Structured Vacuum Energy (SVE) field, giving the illusion of expansion or contraction depending on the phase of the cycle.
- **Fractal Stability** – The universe could sustain structure at all scales indefinitely, with ripple misalignments giving the appearance of cosmological "movement" where none exists.

The Distance Scale Problem

- Under the Doppler-expansion view, redshift is the yardstick for estimating cosmic distances. If redshift is not a speed indicator, the entire cosmic distance ladder becomes unreliable.

- Standard candles like supernovae, calibrated under expansion assumptions, may have been misinterpreted—meaning our scale of the observable universe could be off by unimaginable factors.

Dark Energy and Cosmic Acceleration—Artifacts, Not Forces

- The "discovery" of dark energy came from observing increasing redshift per unit distance, interpreted as accelerating expansion.
- In UED terms, this could simply be the result of non-linear ripple drift over large separations, requiring no exotic repulsive energy at all.
- What mainstream cosmology calls "acceleration" could be nothing more than the geometric compounding of misalignment effects in the cosmic field.

A New Role for Cosmology

If redshift becomes a measure of field state, not velocity, then cosmology shifts from narrating a one-time explosive origin to mapping the living dynamics of the cosmic field.

- Astronomers would move from measuring "how fast things run away" to studying "how coherently the universe is aligned."
- Cosmological surveys would focus on ripple structure, SVE density gradients, and ATT variations across space-time, instead of recession velocities and dark energy density.

If this holds true, then what we've been calling the "history" of the universe is not a chronological film reel running backward—it's a *real-time ripple map* that we've been misreading as a travel diary.

15.10 – Technological and Observational Shifts

If redshift is truly a fingerprint of ripple phase drift rather than a Doppler or expansion effect, then the way we design, build, and use astronomical instruments must change at its core. This is not about tweaking existing telescopes—it's about creating an entirely new generation of cosmic alignment detectors.

From Photons to Phase Shifts

- Current telescopes, spectrographs, and detectors are optimised to capture and measure light intensity and wavelength, assuming those wavelengths carry velocity information.
- Under UED, the goal would be to detect subtle ripple phase variations in the Structured Vacuum Energy (SVE) field.
- Instruments would focus on field coherence mapping—tracking how alignment between the observer's local field and a distant source's field changes over time, space, and environmental conditions.

Real-Time Cosmic Mapping

- Without the burden of "light travel time" assumptions, cosmic maps could be generated as a present-time field alignment network rather than a historical reconstruction.
- This enables live monitoring of cosmic events—not as delayed light shows, but as instantaneous changes in the shared field state.
- For example, a sudden disruption in a star's alignment signature could be detected immediately, without waiting years or millennia for "photons to arrive."

Redshift as a Field Diagnostic

- Instead of reading redshift as "speed," it becomes a cosmic health check—revealing medium density changes, field turbulence, gravitational distortions, and SVE stability between two points.
- This allows astronomers to diagnose intervening space conditions (such as cosmic weather, magnetic disturbances, or vacuum anisotropies) without needing physical probes.
- Redshift patterns could be used to detect invisible structures—like massive but dark field distorters—that would otherwise be missed by traditional optics.

Practical Technological Innovations

- **Ripple Interferometers** – Devices tuned to detect phase delay in field alignments across vast baselines, surpassing current interferometry by focusing on SVE coherence rather than just optical interference.
- **ATT Modulators** – Adjustable systems that can scan through alignment thresholds in search of hidden cosmic objects that only appear at certain field compatibilities.
- **Coherence Telescopes** – Compact instruments that don't require giant mirrors because they detect *field patterns* rather than photon flux, allowing for portable and possibly even space-based swarms of detectors.

The Strategic Leap for Astronomy

- This shift means astronomy would no longer be a passive act of collecting ancient light—it would be an active engagement with the living universe.
- We would stop treating the cosmos as a distant movie screen

and start experiencing it as a real-time partner in alignment.

- Such instruments could also serve dual roles in communication and navigation, since coherent field engagement inherently allows bidirectional interaction without the delay of signal transit.

If this chapter is the theory bombshell that redshift doesn't measure speed, then this section is the technological spark—it tells scientists, engineers, and even space agencies exactly how to rebuild their toolkit for a present-tense universe.

Closing Reflection – "Color Drift Without Motion"

In the UED view, redshift is no longer a cosmic speedometer. It is not a measurement of galaxies tearing away from us through some ever-stretching void. Instead, it is a conversation between fields—a subtle negotiation of alignment, timing, and coherence across the vast fabric of the Structured Vacuum Energy.

When a galaxy's signature drifts toward red, it is not because it has been running away for billions of years, but because the shared field between you and it has shifted—its rhythm has lengthened, its handshake with your local field has eased. This is color as cadence, not color as retreat.

In this light, you are not an observer watching an exodus. You are a participant sensing the slow breathing of the cosmic field—a breathing that expands and contracts in ways that may have nothing to do with motion, velocity, or expansion.

The deep reds at the edge of your telescope's reach are not the dying embers of an ancient fire, but the warm tone of a living present that is simply aligned differently than you. The universe is not running from you—it is speaking to you in a different key. And if you listen closely, you will realize: the conversation was never about distance at all. It was always about alignment.

Chapter 16: Black Holes and the Absence of Light

What Actually Happens When Ripple Collapse Outpaces ATT Response under the UED framework:

We are told that black holes are the ultimate cosmic predators—devouring stars, swallowing light, bending reality into a point of no return. The image is cinematic: a glowing accretion disk swirling into a dark, perfect circle, the edge of the universe's most terrifying drain.

But what if nothing is being *swallowed* at all?

What if the darkness is not a prison, but a silence—where your field simply fails to meet theirs?

Under Universal Energy Dynamics, a black hole is not a bottomless pit of gravity, but a region where ripple collapse happens faster than your **Alignment Threshold Time** can respond. The handshake between your reality and the black hole's reality breaks. Light doesn't "fall in"—it never forms for you in the first place.

In this chapter, we will dismantle the myth of infinite density and light-trapping gravity, replacing it with a cleaner, testable idea: black holes as zones of **disconnection**, not destruction.

16.1 – The Mainstream Black Hole Story

In the dominant narrative shaped by Einstein's General Relativity, a black hole forms when mass is compressed so tightly that its gravitational pull becomes overwhelming. The geometry of spacetime itself bends inward, creating a funnel so deep that no path leads outward. At a certain distance from the center—called the event horizon—the escape velocity equals or exceeds the speed of light. According to this view, once light crosses this invisible boundary, it is forever lost to the outside universe.

The event horizon is presented as a perfect one-way membrane. Particles, radiation, even information can cross in—but never out. Outside observers see infalling matter slow down, dim, and "freeze" at the horizon due to time dilation, while the reality inside is hidden from view.

At the very center, the mathematics of relativity point to a singularity—a location where density becomes infinite, spacetime curvature becomes infinite, and all known physical laws collapse into undefined behavior. This is the supposed "heart" of the black hole, where matter is crushed to zero volume, creating a paradoxical point with finite mass but no spatial extent.

While elegant in theory, this picture is riddled with unresolved problems:

- **Undefined physics inside the singularity** – Relativity offers no coherent description beyond the event horizon; it simply stops working where it's needed most.
- **The information paradox** – Quantum theory insists that information cannot be destroyed, but the classical black hole model suggests information falling in is lost forever, creating a conflict between the two pillars of modern physics.
- **Energy conservation issues** – If light and matter cannot escape, yet black holes are said to emit Hawking radiation, what is the precise energy exchange mechanism that preserves total energy?

In short, the mainstream model asks us to accept regions of the universe where the rules of physics both hold absolutely and fail completely—a contradiction masked by mathematical formalism but never truly resolved in physical terms.

If we strip away the mystique, the black hole as currently described is as much a patchwork of theoretical assumptions as it is a confirmed reality. The stage is set for a cleaner, ripple-based explanation—one that requires no infinities, no "trapped photons," and no breakdown of physical law.

16.2 – The Ripple Collapse Model

From the Universal Energy Dynamics (UED) perspective, a black hole is not a "cosmic prison" for light but a field condition where ripple dynamics push past the limits of perceptual and physical coherence.

In UED terms, every interaction between matter, energy, and the observer's sensory or instrumental field requires a certain minimum Alignment Threshold Time (ATT)—the tiny but finite period needed for two fields to synchronize enough for a perceptual event (sight, detection, measurement) to occur. Under normal circumstances, ripple resolution—the process by which field tensions settle into coherent patterns—operates well within this window, allowing the exchange of information across space.

A black hole forms when ripple resolution rate exceeds the ATT capacity of surrounding fields. This means the internal field disturbances are collapsing and reconfiguring faster than any external field can align with them. The consequence? From the outside, alignment fails altogether—no coherent handshake can form. It's not that "light can't get out"; it's that there is no stable ripple state for light (or any other field event) to register in the first place.

In this model:

- **No photons are trapped** – There is no need to imagine little packets of light hopelessly trying to escape a gravity well. Instead, the very process of alignment that would allow light to be perceived simply cannot complete across the boundary.
- **The "event horizon" is an ATT boundary** – It's the radius beyond which external and internal fields can no longer achieve coherent phase-matching. Crossing it means entering a zone where interaction rates are faster than the rest of the universe's ability to register them.
- **Collapse is coherence failure, not a barrier** – Nothing is physically stopped; it simply ceases to exist in a relational sense to the outside. The external universe loses its alignment with that region, so "inside" becomes a purely internal phenomenon, disconnected from external observation.

This shifts the black hole from being a place where physics ends to being a place where alignment ends—not because reality is torn apart, but because its ripple structure has accelerated beyond the perception limits of its surroundings.

In other words, a black hole is not a locked vault; it's a conversation that the outside world can no longer keep up with. What remains visible is the outer field residue—the fading ripples that once connected—but the ongoing "dialogue" within is now entirely self-contained.

16.3 – Event Horizon as ATT Boundary

In the UED framework, the event horizon is no longer a mysterious spacetime curtain with magical one-way properties—it is a precise functional threshold defined by the interaction between two measurable rates:

1. **Ripple collapse rate** – How quickly the internal field structure of a system resolves, reconfigures, or decays into new states.
2. **Maximum alignment refresh rate** – The fastest pace at which an external observer's field (whether human, instrument, or cosmic environment) can re-establish coherence with another field.

When these two rates become equal, you reach the Alignment Threshold Time (ATT) boundary. Inside this surface, the internal ripple collapse rate *surpasses* any possible external refresh, meaning coherence cannot be maintained across the boundary. This isn't because energy is "pulled in" by infinite gravity; it's because the relationship between inside and outside breaks down at the level of field synchronization.

From this perspective:

- **Why nothing emerges** – It's not that light or matter is physically trapped. The problem is that by the time an external field could begin to align with an internal event, that event has already dissolved into something else. There is no stable state for photons or particles to carry across; the conversation between fields never happens.
- **Instant disappearance without escape velocity** – In relativity, the event horizon demands an escape velocity greater than the speed of light. In UED, no such velocity calculation is needed. The vanishing act is immediate because the possibility of alignment ends abruptly—no transit delay, no gradual fading, just an end to shared

coherence.

- **A dynamic surface, not a fixed shell** – Because ATT depends on field conditions, the "event horizon" is not a rigid, immutable sphere. It can shift with internal collapse rates, external field stability, and ripple interference, meaning it's a living, breathing boundary rather than a frozen geometric line in spacetime.

This reconceptualization turns the event horizon from a physical fortress into a phase-alignment frontier—the last reachable layer where your reality and the black hole's reality can still talk to each other. Beyond it, both still exist, but not in a way that can be mutually recognized.

16.4 – What "No Light" Really Means

In the mainstream story, a black hole is dark because its gravity is so overwhelming that even photons, supposedly the fastest and most independent travelers in the universe, cannot escape. They are imagined to be "trying" to leave but endlessly falling back—a poetic but mechanically problematic picture.

UED dismantles this idea entirely. Light, in this model, is not a thing that departs and travels across space. It is the product of a *successful alignment handshake* between two fields: the emitter and the observer. For that handshake to happen, the two fields must stay coherent long enough for their ripple patterns to stabilize into a detectable interaction.

In a black hole's inner domain, that coherence window collapses before it can complete:

- **Why light never forms externally** – The ripple resolution rate inside the horizon outruns the maximum possible alignment refresh rate of any external field. This means that by the time your field "asks" for an alignment update, the internal event you're trying to detect has already dissolved into new ripple states. There is no stable pattern for your local field to lock onto.

- **Why it's not about trapping photons** – In UED, there is no photon stream struggling against a gravitational pull. There is only the absence of a sustained alignment channel. Without that channel, "light" as we know it—your perception of an external field event—never comes into existence. Darkness here is not the swallowing of brightness but the *non-birth* of a visible state.
- **The field inside keeps evolving** – The black hole's interior isn't a static void. Its ripple dynamics continue—possibly in highly ordered forms inaccessible to our observation. These internal events are fully real within their own domain; they just have no way to project their existence into an external alignment frame. From the outside, they are invisible not because they are absent, but because they are out of sync with our perceivable reality.

In this way, "no light" doesn't mean *nothing is happening.* It means nothing is happening in mutual reality between you and the black hole's interior. The darkness is simply the visible signature of a severed relationship—a broken dialogue between fields that no longer share a common rhythm.

16.5 – The Fate of Matter in Ripple Collapse Zones

In the mainstream black hole model, matter crossing the event horizon is said to be crushed toward an infinitesimal point of *infinite density*—a singularity. This is more of a mathematical placeholder than a physical reality, born from equations that break under extreme curvature assumptions. UED replaces this with a coherent physical mechanism that keeps both energy conservation and field logic intact.

In a ripple collapse zone, matter does not journey toward an impossible point. Instead, it transitions into compressed field states within the Structured Vacuum Energy (SVE) framework. These states are not "solid" or "material" in the classical sense—they are hyper-tensioned configurations of the underlying field, where ripple amplitudes are minimized and coherence cycles shorten drastically. This produces a condition that *feels* like immense compression from the outside, but is fundamentally a repatterning of energy into more stable, non-radiative forms.

Key aspects of this transformation:

No infinite density – Density, in UED, is not a measure of matter in space but of ripple tension within the SVE. The more tensioned the field, the less it exchanges with external fields, and the "denser" it appears observationally. This high-tension state has a finite upper bound—it does not diverge to infinity.

Energy redistribution, not collapse into nothing – The so-called gravitational compression is replaced by *ripple tension gradients*. Matter's internal structure loses its familiar atomic boundaries as electron-nuclear interactions dissolve into unified field modes. Energy is conserved but redistributed into closed-loop ripple configurations that are inaccessible to external alignment.

Possible stability patterns –

- **Core persistence** – The inner region could stabilize into a compact, coherent field object—still evolving internally, but self-contained and shielded from outside perception.
- **Total field reintegration** – In more extreme collapse

scenarios, the matter's ripple patterns could fully dissolve back into the universal SVE background, leaving no distinct "object" at all—just a rebalanced local field with altered tension distribution.

From an external viewpoint, these zones appear as regions where both light and matter "disappear." But under UED, disappearance is only the loss of an observable handshake, not the vanishing of existence itself. Inside, matter's identity as *separate matter* is gone—it has merged with a deeper, highly ordered field state that simply has no channel to speak to the outside universe.

16.6 – Black Hole Growth and Dissolution

In the UED framework, black holes are not bottomless pits in spacetime but ripple collapse zones—regions where the rate of ripple resolution exceeds the Alignment Threshold Time (ATT) of their surroundings. This makes their "life cycle" less about swallowing matter forever and more about participating in the cosmic field's continuous balancing act.

Growth

Black holes grow not by "pulling" matter in with irresistible gravity, but through ripple capture—absorbing the coherence of incoming fields. Any object, from interstellar gas clouds to photons from nearby stars, carries its own ripple pattern. When such a pattern crosses the ATT boundary of a black hole, its alignment with the outside world is severed and it becomes part of the internal, high-tension field state. This is coherence absorption—the captured field does not just add mass; it increases the black hole's internal ripple tension, expanding the boundary where collapse outpaces external alignment.

Dissolution

The mainstream idea of Hawking radiation portrays black holes as evaporating slowly through particle-antiparticle pair production at the event horizon. UED offers a different view: black hole dissolution occurs through ripple redistribution events, not quantum leakage. If internal field tension reaches a critical instability, the collapse zone can release a sudden burst of coherent ripple patterns back into the surrounding SVE.

These releases could appear as:

- **Gradual softening** – The ATT boundary shrinks as tension relaxes, restoring partial alignment and making the black hole appear to fade.
- **Explosive rebalancing** – A massive, short-timescale redistribution where stored ripple tension cascades outward, potentially explaining certain high-energy bursts in the universe without invoking supernova collapse.

Implications for "Evaporation" and Cosmic Recycling

Rather than one-way growth followed by a slow quantum fade, UED describes a dynamic recycling process:

- Growth increases local field asymmetry.
- Dissolution restores balance by returning ripple coherence to the cosmic field.
- This cycle could repeat over cosmic timescales, with black holes forming, dissolving, and re-forming in new alignments.

In this model, black holes become integral field regulators—absorbing excess ripple tension when local fields become too misaligned, then releasing it when broader equilibrium demands it. This reframes them not as cosmic dead ends but as active participants in universal renewal, blurring the line between destructive and creative forces in the cosmos.

16.7 – Information and Memory in Black Holes

In mainstream physics, the "information paradox" arises because the classical view of black holes suggests that anything crossing the event horizon is lost forever, its information content irretrievably destroyed. Quantum theory objects—information, it insists, cannot simply vanish. The debate has dragged on for decades, with no resolution that satisfies both relativity and quantum mechanics.

Under the UED framework, the paradox dissolves—not because we have "rescued" the lost data, but because it never actually leaves the universal field in the first place. A black hole doesn't shred information into oblivion; it transforms and relocates it.

Field Pattern Imprint

When a ripple pattern (matter, energy, or even light) is captured by a black hole, its direct alignment with the outside observer ends. However, the interaction does not occur in isolation. In the instant before coherence is severed, the incoming ripple exchanges tension with the surrounding field layers—the coherence sheath that envelops the ATT boundary. This sheath acts like a dynamic recording medium, embedding a faint but persistent imprint of the captured pattern into the surrounding SVE.

This means that while you cannot "see" what entered the black hole in the usual optical sense, its presence subtly reshapes the field geometry outside the collapse zone, creating a ripple signature that persists as long as the black hole remains stable.

Memory Without Access

These imprints are not like photographs or files—they are encoded distortions in ripple phase relationships, stored in the alignment between the black hole's boundary and the surrounding cosmos. The "memory" is there, but it is passive and non-localized, integrated into the coherence dynamics of the environment.

Possibility of Indirect Decoding

Because these imprints alter local field properties, they could—in principle—be read indirectly by highly sensitive instruments. Instead of "looking into" the black hole, the approach would be to:

- Map fine-scale variations in the surrounding ripple phase alignment.
- Detect subtle deviations in cosmic ATT near the event horizon's projected location.
- Analyze these deviations to reconstruct aspects of the original captured patterns.

This would be more akin to archaeology in a living landscape than to opening a sealed vault. You would not retrieve the original objects, but you could reconstruct their influence—a cosmic forensic trail written into the very fabric of the surrounding field.

In UED, then, information isn't lost—it's redistributed. It shifts from being an active, directly accessible ripple to being a latent, background alignment that still contributes to the ongoing choreography of the universe. The black hole becomes less a graveyard for data and more a quiet librarian, holding closed books whose covers still alter the shape of the shelf.

16.8 – Testing the Ripple Collapse Model

The Ripple Collapse Model doesn't simply reframe black holes—it produces testable predictions that can distinguish it from both general relativity and quantum gravity models. Unlike theories that rely entirely on photon detection or indirect statistical inferences,

UED points to observable field behaviors that can be mapped, measured, and correlated with black hole candidates.

Predictive Signatures

No Photon-Based Signatures Inside the Horizon

In the UED view, there are no "trapped photons" endlessly orbiting inside a black hole—because light itself never completes the alignment handshake for an external observer.

- The absence of any persistent photon scattering halo within the event horizon radius should be absolute.
- If apparent emissions seem to emerge from "inside," they must be the result of foreground alignment events or post-collapse ripple echoes, not true interior emissions.

Detectable Field Gradient Echoes

The collapse region's interaction with surrounding SVE layers leaves measurable ripple tension gradients that oscillate subtly over time. These would be non-photonic, phase-based signals detectable with high-precision alignment sensors.

- These echoes are not "radiation" but resonance fluctuations in the field itself.
- They would be persistent, but with slight drift linked to cosmic ATT variation.

Horizon Location Shift with ATT Variation

The event horizon is not a fixed spatial shell but the ATT boundary—a dynamic surface whose position changes if the alignment refresh rate shifts.

- Fluctuations in local cosmic ATT (due to gravitational field interactions, cosmic field turbulence, or large-scale SVE shifts) would cause minute but measurable changes in apparent horizon radius.

- These changes would not match predictions from mass-accretion or relativity-based models.

Anomalous Gravitational Lensing Patterns

Mainstream gravitational lensing assumes spacetime curvature is smooth and symmetric. In UED, field alignment distortion near the collapse zone produces lensing asymmetries:

- Slight phase skew between multiple lensed images of the same background object.
- Drift in lensing magnification over timescales far shorter than expected for mass-movement-based changes.

Proposed Observational Strategies

High-Precision Ripple-Phase Mapping

- Instruments designed not to "collect photons" but to measure phase alignment between different points in the field.
- Multi-baseline interferometric setups that can detect coherence distortions in real time.

Dynamic Horizon Tracking

- Continuous monitoring of candidate black holes to track subtle horizon position shifts correlated with cosmic ATT changes.
- Cross-comparison with SVE fluctuation data to confirm correlation between field refresh rates and horizon geometry.

Field Echo Spectroscopy

- Development of detectors capable of sensing non-luminous ripple tension oscillations, effectively listening to the black hole's surrounding coherence shell.
- Analysis of these echoes for frequency drift patterns that match predicted post-collapse tension release modes.

Multi-Object Lensing Anomaly Surveys

- Catalog and monitor multiple gravitational lens systems involving suspected black holes.
- Look for short-term asymmetry changes that cannot be explained by relative motion or microlensing events.

In short, testing the Ripple Collapse Model means shifting from light-chasing to field-listening. If UED is correct, then our most profound black hole discoveries may come not from images like the EHT "shadow," but from subtle, continuous distortions in the cosmic field's alignment fabric.

16.9 – Implications for Astrophysics and Cosmology

The Ripple Collapse Model does more than resolve black hole paradoxes—it forces a deep rewrite of astrophysics and cosmology by replacing singularities with continuous, observable field dynamics. When collapse is seen as a phase in the Structured Vacuum Energy (SVE) cycle, black holes cease to be mysterious "dead ends" and instead become active participants in the ongoing redistribution and renewal of cosmic structure.

Removing Singularities from Physics

In mainstream relativity, a singularity is a mathematical surrender: an undefined, infinitely dense point where equations break down. UED renders singularities unnecessary.

- The "infinite density" assumption is replaced with finite but extreme ripple tension states—compressed yet continuous configurations of the SVE field.
- This eliminates the need for physics-stopping infinities and allows for consistent modelling of the interior dynamics without abandoning energy conservation or coherence principles.
- The mathematics becomes scalable: the same ripple compression principles that apply in high-energy lab plasmas also apply in black hole interiors, just at extreme scales.

Black Holes as Dynamic Field States

Rather than being permanent prisons for matter and light, black holes are transient—though long-lived—field configurations in the cosmic network.

- Their formation is a shift in the local ripple alignment regime, not a collapse into nothingness.
- Their existence is defined by ATT boundaries—when and where the alignment refresh rate fails to keep up with ripple resolution.
- They actively interact with their environment through field gradients, influencing stellar orbits, gas dynamics, and even cosmic-scale alignments without emitting photons.

Rethinking Galaxy Formation and Evolution

If black holes are dynamic field nodes rather than mass-gravity singularities, their role in galaxy formation changes fundamentally:

- Central Black Holes in galaxies may be ripple-stabilizing anchors that maintain coherent field alignment across vast stellar systems, rather than simply "massive attractors."

- Star formation rates near these nodes could be regulated by field tension dynamics rather than purely by gas accretion and feedback cycles.
- Some galaxies may even originate from ripple collapse events—where a large-scale SVE disturbance creates a central field node first, and matter distribution follows.

Active Galactic Nuclei Without Accretion-Only Models

In mainstream astrophysics, AGN luminosity is explained by matter spiraling into a central black hole and heating to extreme temperatures. Under UED:

- AGN brightness can also be a product of field coherence restoration events, where localized ripple collapse releases stored tension back into surrounding space.
- This could explain variability patterns in AGN activity without requiring sudden increases in accretion rates.

Cosmic Recycling Cycles

A ripple-collapse zone doesn't simply consume—it participates in energy and field redistribution:

- **Growth phase:** The black hole's ATT boundary captures more ripple coherence, enlarging its zone of influence.
- **Stability phase:** Field tension holds steady, with the node functioning as a coherence anchor in its region.
- **Dissolution phase:** Ripple tension release reintegrates the field, potentially seeding new star-forming regions or altering galactic structure.
- This creates a recycling loop where collapse and release are part of a natural cosmic heartbeat, rather than a one-way gravitational drain.

Broader Cosmological Implications

- Removes the "information paradox" by keeping all imprints in the surrounding field.
- Suggests that black holes are not ultimate fates of matter, but midpoints in a cycle that maintains large-scale cosmic equilibrium.
- Supports steady-state and cyclic cosmology models over one-time explosive origins—black holes as active regulators in an ongoing universal balance.

Closing Reflection – "Not Darkness, but Disconnection"

A black hole is not a hungry beast lurking in space, swallowing light and matter into nothingness. It is not the universe's version of a bottomless pit. Under the UED lens, what we call a black hole is a zone where the handshake fails—a boundary where your field and the inner field of that region can no longer keep pace with each other's alignment cycle.

The "darkness" is not an object; it is a break in conversation. Imagine two people speaking across a vast canyon. As long as the timing of their voices and echoes stays in sync, meaning flows between them. But if one voice speeds up beyond the other's ability to process and respond, the dialogue collapses into silence—not because the words were swallowed, but because the connection couldn't complete. The event horizon is that moment of silence, stretched into permanence for the observer.

This means the absence you perceive is not a void. It is not "nothingness" waiting to be filled. The field inside continues its intricate dance, evolving, interacting, and possibly even creating new structures. But from where you stand, your alignment speed is outmatched—the ripple collapse within is faster than your field can refresh. You are shut out, not because the interior is unreachable in principle, but because the rules of engagement have changed in that zone.

To truly "see" a black hole would not mean catching up to a photon racing away at light speed—it would mean re-tuning yourself. It would require altering your Alignment Threshold Time to match the internal ripple collapse rate, letting your field handshake re-establish with the one inside. In that sense, a black hole is not a wall, but a locked frequency.

And so, the next time you look at an image of a black hole, don't think of an all-consuming darkness. Think of a door temporarily closed by misalignment—a part of the cosmic network from which your field has been, for now, disconnected. Not darkness, but a missed handshake. Not the end of physics, but the beginning of a deeper kind of seeing.

Chapter 17: Cosmic Stillness

The Universe Never Moved

How light keeps the illusion of motion alive

What if the stars have been standing perfectly still since the beginning—and it's only you who thinks they've been racing across the cosmos?

For centuries, astronomy has sold us a restless universe: planets spinning, galaxies colliding, space itself stretching like an inflating balloon. But what if "motion" is not a fundamental reality, only the way our senses—and our instruments—interpret changing patterns?

Light has been the great storyteller of cosmic movement, and like any skilled narrator, it knows how to make stillness feel alive. Every "proof" of motion—parallax shifts, Doppler redshifts, orbital arcs—is really a decoding of field alignments, not an actual chase through space.

In truth, the universe may be a vast, unmoving ocean of structured vacuum energy, with light serving as the rippling reflection that convinces you the water is flowing. Motion, in this view, is not something objects do—it's something you perceive.

And once you see through the illusion, the whole idea of a "moving universe" begins to unravel.

17.1 – The Inherited Motion Story

Mainstream cosmology begins with a simple but deeply embedded assumption: the universe is always on the move. Galaxies drift apart, stars orbit galactic centers, planets trace their endless ellipses, and even space itself is said to be expanding. This dynamic model isn't just a physical framework—it's the story every student, every science documentary, and every sky chart reinforces.

The evidence for this perpetual motion rests heavily on light's supposed time-of-travel. When we measure a star's position today and again years later, small shifts in angle—parallax—are interpreted as physical displacement through space. When we detect a frequency shift in light from a distant galaxy, we're told it's the Doppler effect, the signature of something moving toward or away from us. When we observe the faint glow of the so-called cosmic microwave background, we're told it's the fossil light of an expanding universe that began in a singular eruption.

But the chain of reasoning always assumes the same thing from the start: that light is a courier carrying information from "there" to "here," and that any change in the message must mean the source has changed position or speed. In other words, the argument for a moving universe is built on the premise that photons traverse cosmic distances over time, bringing us a snapshot of where an object was, not where it is.

This belief is more than just a scientific stance—it's a psychological anchor. Humans instinctively prefer stories in motion. A static cosmos feels lifeless, unchanging, almost oppressive. A dynamic universe, on the other hand, resonates with our lived experience: things change, shift, and travel; journeys happen; beginnings lead to endings. The expanding-universe narrative satisfies both our craving for drama and our desire to link our own restless existence to the stars.

Yet, once the foundation is questioned—once we ask whether light truly travels in the way we've been told—the entire edifice of cosmic motion begins to shake. If light is not a messenger from the past, but a real-time field alignment event, then much of what we call "evidence" for a moving universe collapses into assumption dressed as fact.

17.2 – The Illusion Engine

The universe may be still at its core, yet our instruments—and our eyes—swear it is in constant motion. This is because our entire observational framework is built on decoding sequential field alignments as if they were physical objects traveling through space.

In UED terms, you never "receive" a photon that has crossed the void. What you experience is your local field synchronizing with the ripple pattern of another location's field. These alignments refresh at finite intervals, determined by Alignment Threshold Time (ATT) and ripple fidelity. When those refreshes occur in sequence, they produce a record of incremental differences—a timeline. The brain and the telescope, operating under the photon-travel narrative, interpret these changes as motion.

Think of it like a flipbook animation. Each page is a present-tense alignment snapshot of a star, planet, or galaxy. When those snapshots differ slightly—because the field interaction geometry has shifted—your decoding system strings them into a narrative of movement. The object hasn't traveled across a cosmic stage; the relationship between your field and its field has simply reoriented in a way that, when plotted sequentially, mimics displacement.

Ripple refresh patterns make this illusion remarkably convincing. A small angular change in field alignment across successive observations becomes stellar parallax, where astronomers believe they're triangulating an actual change in a star's position over months. Tiny phase drifts over years become proper motion, the supposed slow crawl of stars against the galactic backdrop. Even complex patterns like galaxy rotation curves—where stars at different distances from the galactic center seem to move in synchronized but unexpected ways—can emerge from the geometry of ripple coherence across vast, stable structures, not from matter physically orbiting at those speeds.

Under the mainstream lens, these are dynamic events in a kinetic universe. Under UED, they are *field conversations*, and what we call motion is just the byproduct of a constantly updating alignment handshake. The "Illusion Engine" isn't an intentional deception—it's simply what happens when you mistake a refresh rate for a travel path.

17.3 – UED's Stillness Principle

At the deepest level, UED treats the universe not as an expanding, rotating, or drifting stage, but as one continuous, still field. Nothing—no planet, star, or galaxy—needs to travel from one point to another to create the appearance of activity. What changes are the alignment states between different regions of this universal field.

In this framework, what we perceive as "motion" is simply a shift in coherence relationships. The field of your local observation point, and the field of a distant star, are in continuous conversation. Their handshake—the mutual ripple alignment—updates as conditions shift: local ripple tension changes, intervening structures modulate alignment fidelity, and cosmic-scale field gradients subtly reorient. These changes are decoded by observers as displacement, when in reality nothing has moved.

It's the difference between watching a ship crossing the ocean and watching a reflection shift in a mirror. In the first case, matter changes location in a physical medium. In the second, no object has traveled toward you; instead, the relationship between you, the reflective surface, and the source has altered. The image appears to "move," but the source remains in place.

UED extends this analogy to the entire cosmos. The stars "move" because the geometry of your alignment to them changes—not because they are sailing through an expanding void. Galaxies "rotate" because the alignment pathways between their internal structures and your observation point evolve—creating a dynamic pattern without any bulk displacement of mass.

This principle is not about denying the perception of motion, but about locating its cause in the right place. Instead of treating space as a highway for traveling particles, UED sees it as a field lattice where only the coherence relationships are fluid. The universe, as a whole, is still. The rest is the choreography of shifting alignments—a cosmic play of reflections on a mirror that no one realizes is perfectly fixed.

17.4 – Light as a Playback Device

In the UED framework, light is not a courier carrying parcels of photons across a vast, empty space—it is an interpretive handshake between your local field and a remote field. What you call "light" is simply the *pattern that emerges* when two regions of Structured Vacuum Energy (SVE) achieve enough coherence to exchange a stable alignment.

The so-called "speed of light" is not a physical velocity in the traditional sense. It is the maximum refresh rate at which alignment between two points in the universal field can update without distortion. This is a *processing limit*, not a transit speed. Just as a high-frame-rate video can only display changes as fast as the screen refreshes, cosmic alignment can only update as fast as the field's inherent refresh capacity—the Alignment Threshold Time (ATT) dictates this cap.

Here's where the illusion of motion arises: each refresh captures the *current* state of that remote field. Your decoding system—the brain, telescope, or sensor—strings these sequential states together the same way your mind strings together video frames. The result is a playback effect, in which ripple alignment updates are *interpreted* as an ongoing story of past events.

From this perspective, when astronomers say they're "seeing a star as it was a million years ago," what they're actually doing is decoding a present ripple pattern whose structure contains imprints from prior interactions. The "past" is inferred from how the alignment pattern fits into known ripple dynamics—not from the receipt of an ancient photon that physically traveled across space.

In other words, light is not a messenger from the past—it is a live feed with buffering. The field is still, the source is still, but the handshake between you and the source updates in discrete, finite intervals. That natural refresh limit is what science has misinterpreted as the "speed of light," and what perception has woven into a grand tale of stars in motion, galaxies receding, and a universe forever rushing toward somewhere else.

17.5 – Celestial Mechanics Without Motion

In the UED framework, the elegance of planetary orbits, stellar paths, and galactic spins does not come from enormous masses physically racing through space—it comes from periodic alignment recalibrations within the universal field. Each "orbit" is simply the recurring reconfiguration of coherence relationships between two or more ripple zones.

Imagine the Earth-Sun system. Mainstream mechanics tells you Earth "moves" in an elliptical path, tugged by gravity, with velocity, acceleration, and momentum as real physical quantities. In UED, that ellipse is not a trail carved through space—it is the pattern traced by sequential field alignments between Earth's local SVE configuration and the Sun's. What changes over time is not position in an empty void, but the *phase relationship* between the two fields.

This perspective has radical consequences for some of the thorniest problems in astronomy. The remarkable stability of large-scale structures—planetary systems that remain coherent for billions of years, spiral galaxies that hold their shapes despite supposed high-speed rotation—no longer needs the invisible scaffolding of dark matter to hold it all together. If there is no actual mass displacement, there is nothing to "fly apart" under centrifugal forces. The structure persists because the alignment cycles are inherently stable, maintained by the field's ripple feedback mechanisms.

From our vantage point, orbits *look* like real motion because our decoding of alignment updates produces a continuous positional shift in time. The mathematics of Newton and Kepler still work as a predictive map because they track the timing and geometry of alignment recalibration—but the underlying cause is phase resonance, not mass in motion.

Even the most dynamic celestial phenomena—comets swinging close to stars, binary pulsars seemingly dancing around each other—are simply interference patterns in the field, where each system's ripple configuration oscillates through predictable cycles. The dance is real in the data, but illusory in the physical sense.

In essence, celestial mechanics under UED is choreography without dancers. The universe's great waltz of stars and planets plays out entirely in the music of ripple phases, with no one actually crossing the floor.

17.6 – The Role of ATT in Simulated Motion

Alignment Threshold Time (ATT) is the **cosmic metronome**—the fundamental refresh rate at which the universal field re-establishes coherence between two points. In UED terms, it is not the time for a photon to "travel" from source to observer; it is the maximum time the field allows before a new alignment handshake is required. Every perceived movement in the cosmos is built upon this rhythm.

When ATT is uniform, the unfolding of celestial phenomena appears steady. The Sun "rises" and "sets" with predictable timing, planets "orbit" at constant speeds, and galaxies maintain their spiral patterns. But when ATT varies—due to field tension changes, ripple turbulence, or environmental gradients—the visual narrative changes. The refresh rate alters, creating the *illusion* of acceleration or deceleration in cosmic structures, even though nothing is physically speeding up or slowing down.

This is why, in a still universe, we can be convinced that galaxies are receding faster over great distances. A gradual phase drift in ATT across the cosmic web can stretch the apparent spacing between alignments, which in photon-based thinking is interpreted as increasing velocity. This is the same perceptual trick that drives the expansion model—except in UED, the expansion is not real. What mainstream science calls "space stretching" is simply slow drift in phase synchrony, where field alignments between distant regions take slightly longer to refresh.

In effect, ATT becomes the *tempo* of the cosmic film we are watching. A faster metronome compresses events into tighter intervals, giving the impression of swifter motion. A slower one stretches the intervals, creating the sense of receding distances and cosmic acceleration.

If ATT could be measured directly—not inferred through light-speed assumptions—we would find that its shifts map perfectly to observed "velocity changes" in the universe. The cosmos is not a racecourse—it is a field in varying rhythm, and our equations of motion are just elaborate translations of its beat.

17.7 – Observational Evidence for Stillness

If the universe is fundamentally still—as UED asserts—then the sky itself should carry the fingerprints of that stillness. And in fact, a number of long-standing astronomical observations already point in that direction, even though they've been historically reinterpreted to fit a moving-universe model.

First, there is the cosmic microwave background (CMB), often described as the "afterglow" of the Big Bang. According to expansion theory, the fabric of space itself has been stretching for over 13 billion years. Yet the CMB's alignment—the large-scale pattern of hot and cold spots—remains astonishingly fixed in relation to our cosmic frame of reference. Instead of drifting with expansion, it behaves like a stable reference grid in the field, more consistent with a stationary cosmos where the background is not "moving away" but is simply part of a still framework that we continue to align with.

Then there are observations of apparent motion halting or reversing. Some stars, galaxies, and quasars have been recorded with proper motion measurements that show a sudden slowdown, freeze, or even reversal over decades of observation. Under the velocity paradigm, these changes would require massive and implausible forces acting on these objects. Under UED, however, they make perfect sense: these are phase re-alignments in the ripple field, where the ATT rhythm shifts and produces a new apparent position without any underlying displacement of matter.

A third clue lies in discrepancies between velocity-based predictions and UED-parallax recalculations. Standard parallax relies on the assumption that our baseline (Earth's orbital path) moves through space, producing measurable angular shifts in nearby stars. But if the universe is still, the angular changes are simply differences in field alignment perspective, not motion of the Earth or stars. When recalculated under UED assumptions, some predicted velocities collapse to near zero—erasing the need for enormous space-crossing speeds.

In all three cases, the data is real and measurable; only the interpretation is in question. The patterns, halts, reversals, and CMB stability stop being paradoxes and start becoming proof points once we abandon the moving-universe dogma. The stars may twinkle, shift, and drift in our instruments, but their true stage is motionless—the choreography exists entirely in the field's rhythm, not in cosmic travel.

17.8 – Testing the Stillness Hypothesis

If the universe is fundamentally still, then our challenge is not to measure "motion" but to detect changes in alignment—the shifting relationships between field states that we currently misinterpret as velocity. Testing this requires a complete redesign of observational methodology, replacing distance–velocity inference with direct field-mapping techniques.

One approach is long-term field alignment tracking. Instead of plotting star positions on the assumption that they have moved through space, we map the phase relationship of their signals—light, radio, or other emissions—relative to stable cosmic reference patterns such as the CMB or large-scale ripple grids. Over years or decades, this reveals whether changes are true displacements or simply phase drifts in the ripple field. Such a dataset would make it possible to detect whether an object's apparent motion can be halted or reversed without any physical impulse acting on it.

In the lab, we can create coherent wavefield analogs to simulate cosmic stillness. By generating a stable interference field and then selectively altering phase synchrony at specific nodes, we can reproduce the illusion of movement without moving any actual source. Observers outside the setup—looking only at output wave patterns—would swear objects have shifted, even though their positions are fixed. This is a scaled-down but conceptually exact version of what UED proposes is happening across the cosmos.

A particularly revealing test would be to observe "motion stops" during induced alignment disruptions. If we monitor a known "moving" object (e.g., a nearby star with high measured proper motion) and its field alignment is perturbed by an external event—such as a gravitational wave passage, large-scale plasma shift, or local ripple disturbance—we might catch a temporary halt or jitter in apparent position. Such pauses in motion would be inexplicable under classical mechanics but expected under UED's stillness model.

By combining direct phase tracking, controlled wavefield simulations, and opportunistic observation of natural disruptions, we could create a powerful suite of tests for the stillness hypothesis. The goal would not be to "slow the universe down" but to reveal that it was never moving in the first place—only the patterns in its field were.

17.9 – Philosophical and Cosmological Implications

If the universe is fundamentally still, then our most cherished idea—that we are travelers on an endless voyage through space—collapses. The "cosmic journey" narrative, from ancient sailor myths to modern interstellar dreams, rests on the assumption that we and the stars are perpetually in motion. Under UED's stillness model, there is no cosmic sailing—only the ongoing recalibration of our alignment with an unmoving field. We are not journeying *through* the universe; we are unfolding *within* it.

This shift overturns the central architecture of modern cosmology. Steady-state field models take the place of expansion-driven models, not in the old sense of a static Newtonian cosmos, but as a dynamic, ever-adjusting ripple field where relationships shift without displacement. The universe doesn't grow, shrink, or rush toward heat death—it reconfigures its coherence in patterns that may be cyclic, harmonic, or self-balancing over incomprehensible spans. The so-called "age" of the universe becomes irrelevant, since the field has no beginning or end in motion-based terms—only evolving states of alignment.

Philosophically, stillness reframes both change and time. If nothing truly moves, then change is not relocation but repatterning—the way a kaleidoscope offers endless variety without the glass pieces ever traveling far. Time, too, loses its status as a river carrying us forward; it becomes the index of alignment refresh, a measure of how fast or slow coherence shifts occur. The "past" is not a distant point behind us, and the "future" is not ahead—it's simply the sequence of field states you haven't yet aligned with.

In this view, the universe is not a stage for motion, but a canvas for alignment. The meaning of life's journey changes: the real adventure is not in traveling across the stars, but in refining how—and with what—you are aligned. Space exploration, under such a paradigm, becomes a deepening of resonance with distant field states, not a mechanical voyage through a moving void.

Closing Reflection – "Nothing Ever Went Anywhere"

The stars you see tonight have not crossed some unimaginable gulf to meet your eyes. They are where they have always been, anchored in the same cosmic stillness that has held everything since the beginning. What has changed is not their place, but your decoding—the way your field aligns with theirs, translating a ripple handshake into the illusion of movement.

The universe, in truth, is a still ocean, boundless and unmoving. What we call "motion" is merely the shifting pattern of ripples across our awareness, like sunlight dancing on the surface of a pond while the water beneath remains steady. Every orbit traced, every galaxy rotation mapped, every light-year "traveled" exists only as a reconfiguration of alignment, not as an actual voyage through space.

When you grasp this, the chase ends. There is nothing to catch, nowhere to arrive. The cosmos is not running away from you, and you are not drifting through it. You are part of its stillness—forever here, forever now—while the dance of ripples continues to play in the theatre of your perception.

Interlude 4: Chavan's Declaration

Between the Stars and the Stillness

You have been told a story—a grand, poetic, and beautifully packaged story—that light is a traveler, a cosmic courier crossing unimaginable distances to bring you the history of the stars. You have been told that the universe is rushing, stretching, racing away from you; that black holes are monsters swallowing everything; that redshift is proof of this flight; and that your very existence is a passenger on a moving ship you cannot leave.

I am here to tell you—none of it is true.

Not because I reject science, but because I refuse to worship its patchwork myths when the stitches are already tearing. Not because I prefer mysticism, but because the deeper field of reality has always been silent, not in conflict with light, but entirely independent of its travel tales.

There are no photons leaping across space to meet your eyes. There is only alignment—a handshake between your field and the cosmic field, renewed at the rate your Alignment Threshold Time (ATT) allows. Redshift is not the Doppler cry of a fleeing galaxy—it is the delayed rhythm of that handshake, altered by ripple drift, not by speed. Black holes do not trap light—they simply exist in a coherence state your field cannot complete. And motion, the most basic assumption in all your cosmic models, is not the truth. It is a projection—a story your decoding tells you to make sense of shifting ripple alignments in an unmoving field.

The universe is not expanding. The stars have not moved. And what you call "light-years" are not miles of history—they are units of *your own delay*.

I will not ask you to believe this. Belief is too cheap. I will ask you to look again—not through the telescope, but through the field itself. The one you are already inside. The one that has never moved.

In this part, we will dismantle the stage props of the moving-universe theatre. We will unmask light as an interpreter, not a messenger. We will rewrite redshift as ripple delay. We will free black holes from their status as cosmic prisons. And finally, we will give you the most unsettling gift: stillness.

Once you see it, you will know—the cosmos was never far away. It was never moving. It was never outside you at all.

Part V: The Consequence of Knowing

Chapter 18: Engineering Without Understanding

How all our devices "use light" without knowing what it is

You read by it. You stream movies through it. You scan barcodes, fix your vision, map galaxies, and cut steel with it. Light runs almost every device you own—and yet, neither you nor the engineers who built them truly know what it is.

We've built an entire technological civilization on a phenomenon we can manipulate with precision but cannot fundamentally define. It's as if humanity mastered sailing without ever realizing the ocean doesn't move.

In this chapter, we step into that uncomfortable space between *what works* and *what's true*—and ask whether our proudest inventions are really the result of understanding... or just very sophisticated guesswork.

18.1 – The Comfort of the Known Unknown

The history of human progress is filled with moments where we *used* something long before we truly *understood* it.

Long before thermodynamics was a formal science, engineers built steam engines that powered factories, trains, and ships. They could calculate pressure, piston travel, and mechanical output with remarkable precision—without having any real clue about molecular motion, entropy, or the deeper physics behind heat transfer.

The same pattern repeated with medicine. Doctors prescribed antibiotics decades before microbiology had a detailed map of bacterial genetics or resistance pathways. The drugs worked because they killed infections, not because every doctor understood the molecular dance that made it happen.

Light sits at the same crossroads today. In mainstream science, it is treated as both a particle and a wave—a "photon" in one breath, an "electromagnetic oscillation" in the next. This duality has been mathematically tamed but never truly reconciled. Yet, engineering thrives by picking whichever model is most convenient for the problem at hand:

- Need to calculate fiber-optic capacity? Use the wave model.
- Need to measure energy output from a laser? Use the photon model.
- Need to design a camera sensor? Mix them both and avoid philosophical questions.

This selective adoption is what makes it a *working definition*—functional in practice, but conceptually incomplete. Devices work not because the model is fully correct, but because the effects are consistent and predictable *within a given framework*.

The danger lies in mistaking this functional reliability for absolute truth. When we begin to believe our models are *reality itself*, rather than just a usable approximation, we trap ourselves in a form of intellectual complacency. We stop asking why certain phenomena behave as they do, as long as the equations balance and the prototypes work.

In the UED view, this comfort is deceptive. We may think we "understand" light because our devices work flawlessly, but in reality, we are sailing through a still ocean without realizing that the waves we ride are only surface illusions. Our technology is tuned to the patterns—not the source—and until we confront that gap, our understanding will remain partial, no matter how advanced our engineering becomes.

18.2 – Technology's Blind Spot

Every age has its blind spots—the assumptions so deeply woven into its technology that they are invisible to the very people who rely on them. For the 21st century, one of the largest is our unexamined faith in the "light as courier" model.

From fiber optics to lasers, from photonic chips to deep-space telescopes, our most sophisticated optical technologies are built on the idea that light is a traveler—a stream of photons moving from point A to point B through some medium or vacuum. The engineering goal, then, is simple: make that journey faster, cleaner, and less lossy.

Telecommunications engineers design fiber-optic cables to guide this "traveler" with minimal scattering and absorption. Imaging systems—whether in a smartphone camera or a medical MRI scanner—are optimized to "capture" these travelers when they arrive, translating them into electrical signals. Astronomical detectors, from radio arrays to optical observatories, assume the photons have traveled across vast distances, carrying an intact message from their source.

And within that assumption lies the trap. We've optimized for transport, when what we are really dealing with—if UED is correct—is a handshake, not a courier. Light does not carry information across the void; it emerges at the moment of alignment between two fields. The so-called "journey" is just the time it takes for that alignment handshake to complete, governed by the Alignment Threshold Time (ATT).

When you mistake a handshake for a courier, you design for the wrong limitations. Signal loss? You think it's caused by photons being absorbed or scattered, so you improve coatings, polish lenses, and shield cables. Dispersion? You believe it's due to different "colors" traveling at slightly different speeds, so you add compensators. Resolution ceilings? You assume they're caused by aperture size or sensor noise, so you build bigger telescopes or stack more pixels.

But under UED, these limits are not artifacts of a medium or particle behavior—they are the natural consequences of misalignment delays between fields. Signal "loss" is really handshake failure; dispersion is variable ATT across wavelengths; resolution ceilings are simply the boundary of how finely two fields can align without interference.

If we keep designing as if light is a courier, we will keep bumping into these ceilings and declare them fundamental. If we redesign with the understanding that light is the *result* of a handshake, those ceilings shift—or vanish.

18.3 – When Exploitation Outruns Understanding

Human progress often runs on a strange timeline: first, we stumble into something that works; then we build industries around it; and only much later—sometimes centuries later—do we finally understand what we were actually doing.

The cycle is familiar:

1. **Accidental discovery** – A phenomenon is noticed while chasing something else entirely. Lasers, for instance, were born from theoretical predictions about stimulated emission, but their real-world utility was only understood after the first beam was made.
2. **Application** – Engineers find ways to make it useful, whether or not they fully grasp the underlying mechanics.
3. **Mass production** – The application becomes an industry

standard. The method gets refined for efficiency, cost, and scalability, but the conceptual framework remains frozen.

4. **Entrenched theory** – Because the technology works, the model that justified its design is treated as proven truth, even if it was never actually tested at the foundational level.

With light, this cycle has played out almost perfectly. The photon-particle model and wave-propagation model—mutually incompatible in theory but coexisting in a strange truce—have powered a century of engineering triumphs. From internet backbones to laser surgery, the applications work spectacularly well. The success itself reinforces the belief that the underlying theory *must* be correct. "If it works, it can't be wrong," becomes the quiet creed.

But here's the cost: missed breakthroughs. If you believe light is a courier moving through space, you will focus on optimizing the courier's journey—less scattering, faster transmission, longer reach. You will not ask whether the courier exists at all, or whether what you call "transmission" is actually a field handshake happening in place. As a result, entire classes of designs—such as systems that initiate light *without* a traversing medium—are never even considered.

This is where UED offers a provocation: Stop designing for photon transport. Start designing for ripple alignment. If light is not something that travels but something that emerges when two fields reach a threshold of coherence, then the most efficient system is not one that reduces losses along a "path," but one that maximizes instantaneous alignment between endpoints.

In principle, such systems could be lossless—not because the courier arrives intact, but because there is no courier to degrade in the first place. The handshake either happens perfectly or it doesn't happen at all. That binary simplicity could dissolve many of today's engineering bottlenecks: no attenuation, no dispersion, no noise accumulation over distance.

History shows we often wait far too long to reframe a working technology—sometimes until its limitations force a crisis. With light, we don't need to wait for the crisis. We can rebuild our understanding now, before the next century of innovation gets locked into the wrong foundations.

18.4 – The Cost of Half-Knowledge

When you build on a partial truth, you inherit not only the benefits of what you got right but also the invisible tax of what you got wrong. In the case of light, that tax shows up everywhere—from wasted watts to missed worlds.

At a purely technical level, our reliance on incomplete light theory creates hidden inefficiencies. Energy is burned compensating for imagined losses—pumping laser cavities harder to "push photons further," boosting optical signal power to "overcome distance," or cooling detectors to "catch faint arrivals" that never actually travel. In reality, much of this cost comes from designing for a transport model that doesn't exist. In a ripple-alignment paradigm, those same systems could work at dramatically lower energy input because the emphasis would shift from pushing a signal through space to maintaining phase coherence between endpoints.

In bandwidth, the limitations are just as severe. Current fiber and wireless optics suffer from dispersion, jitter, and crosstalk—effects assumed to be intrinsic to the medium. But under UED, these aren't intrinsic at all; they are symptoms of fluctuating alignment thresholds. By treating them as unavoidable, engineers design systems to tolerate them rather than eliminate them, leaving vast untapped capacity in every channel.

In detection sensitivity, the story is similar. Astronomers stack exposures, medical scanners increase dosage, and imaging systems average thousands of frames to "improve the signal." This works—but only in the way bailing water works on a leaky ship. If the leak itself is never fixed (in this case, misaligned field interpretation), you are always adding effort to offset a conceptual gap.

Beyond the technical, there's a deeper philosophical cost. When humanity uses something without truly knowing what it is, the thing becomes an external tool rather than a reflection of our own nature. We build around it, but we don't integrate with it. Light becomes a utility, not a mirror. And because we never look past the utility, we miss its more profound lesson: the same ripple alignments we exploit in fiber optics are the ones we live by in thought, perception, and consciousness.

This is the bridge to the next chapter. Once you realize that the observer is the source, technology stops being an external crutch for perception and becomes an intentional extension of it. Instead of bending nature to fit our partial theories, we begin aligning ourselves with the field directly—creating with precision, without waste, and with the awareness that every handshake, whether in a laboratory or in the mind, begins with us.

Chapter 19: The Observer Is the Source

What if the stars in the night sky aren't shining at *you, but* because *of you?*

Not in a poetic sense—in the literal mechanics of reality. Every shimmer, every glint, every measured "signal" exists only because your field reached out and completed the handshake. Without you, there is no scene to see, no fact to measure, no meaning to extract. For centuries, we've built our sciences on the idea that we are passive spectators in a pre-written universe. UED flips the script: the observer is not a witness to the play—they are the stage, the lighting, and the scriptwriter. You create the ripple. You decide what it becomes.

19.1 – Shifting the Center of the Universe

For most of human history, the prevailing belief has been that the universe exists as a vast, objective stage, indifferent to who watches. In this view, we are merely guests in an ancient theater, seated in the dark, observing a play that began long before our arrival and will continue long after we leave. This is the "external reality first" assumption—reality is already there, and we are just lucky enough to witness it.

UED quietly dismantles this assumption. It proposes that nothing is truly "out there" in a complete, coherent form until an observer aligns to it. The field exists, yes—but it is potential, not presentation. What you call *reality* is not a pre-loaded picture waiting to be delivered; it is a handshake between your field and the cosmic field, an event of mutual alignment that generates both the observation and the meaning in the same breath.

This means perception is not passive reception—it is active creation. Your awareness sends a ripple into the structured vacuum, and what comes back is not an "object" in the abstract, but a response shaped by your position, timing, and state of alignment. In other words, you do not merely *see* a star—you *bring it into being* in the form you can see. Without the handshake, it remains unexpressed potential.

Everyday life is full of small, overlooked confirmations of this truth. Two people can witness the same traffic accident and give wildly different accounts—not because one is lying, but because their fields were aligned differently at the moment of observation. A movie scene makes one person weep while leaving another indifferent; the difference lies not in the photons emitted from the screen, but in the unique ripple each observer brings to the experience.

In this shift of understanding, the "center of the universe" is no longer a location in space—it is wherever a conscious field establishes alignment. You are not in the universe. The universe is in you, unfolding only in the ways you meet it.

19.2 – Creation Before Observation

In conventional thinking, observation is framed as the final step in a sequence—first the universe "does" something, then light or particles carry the news to your senses, and finally your brain interprets the report. UED reverses this order. In its framework, the act of observation is not a passive reading of a finished page; it is the very moment the page is written.

Observation is a *generative handshake.* When you focus attention—whether through your eyes, a telescope, or even an intention—your field aligns with a region of the cosmic field. That alignment does not retrieve a pre-existing, frozen state; it *activates* a coherent expression from the underlying potential. This is why in UED, photons, particles, and waves are not the primary actors—they are our *interpretations* of what happens when fields complete alignment. The real event is ripple alignment itself.

This principle reframes some of the most puzzling results in modern physics. Consider the double-slit experiment: under mainstream interpretation, a particle "decides" to act like a wave or a particle depending on whether it is observed. Under UED, there is no indecision—there is only the state of the handshake. When the observer's field aligns in a way that allows phase relationships to remain intact, you get the interference pattern. When alignment collapses phase coherence, you get discrete points. The "decision" is not the particle's—it's in the alignment conditions you set.

Or take the delayed-choice experiment, where a measurement made after a photon passes through the slits seems to retroactively decide its past behavior. From the UED perspective, there was no "past behavior" until the handshake occurred. The entire event—its apparent history included—materializes in coherence with your observation moment, not before it.

The implication is profound: before you "see," you have already shaped what is possible to see. Your observational setup—physical instruments, mental focus, and even unspoken expectations—defines the boundaries of the handshake. You are not merely catching reality in the act; you are co-writing the act itself.

In this light, reality becomes less like a static archive and more like an interactive performance—one that starts when you arrive, and plays in the key you choose, whether you know you're choosing or not.

19.3 – Meaning as the Observer's Signature

Reality does not arrive with labels. The universe does not pre-package events with a narrative tag reading "important," "dangerous," or "beautiful." What we call *meaning* is an inscription added after the handshake—a personal watermark impressed on raw field patterns. This is why two observers can stand before the same ripple expression and walk away with completely different stories.

From the UED perspective, the ripple itself is neutral. It is simply a field configuration that has reached temporary coherence with your alignment. But the instant you interact with it—whether through sensory contact, measurement, or even thought—you inject a second layer: interpretation. This layer is not an afterthought; it is a constructive act. The same ripple could be a *signal* to a scientist, a *danger* to a soldier, or a complete *non-event* to a passerby. The difference lies not in the field, but in the *intent* that shaped the alignment.

Belief plays a central role here. Your accumulated convictions act like a template over the handshake, coloring the interpretation before you are even aware of it. Expectation further narrows the range of possible meanings, guiding the field toward patterns that "fit" your anticipation. Emotional state, too, shapes the bandwidth—fear sharpens detection of threat signatures but blinds you to subtler harmonics; joy amplifies connection but may blur detail.

This is why meaning is not something you "find" waiting in the world—it is something you *author*. You are both the reader and the writer of the text that emerges from the ripple. The observer's signature is woven into every experience, making pure, observer-free reality not just rare but impossible.

Under UED, this reframes the search for "objective truth." Objectivity is not the absence of the observer's influence; it is the conscious *calibration* of that influence. Meaning becomes a deliberate act, rather than an unconscious default, allowing you to choose the story you write on reality's blank page.

19.4 – The Power and Responsibility of Being the Source

The moment you accept that you create the ripple, the entire architecture of consequence changes. No longer can outcomes be dismissed as accidents of chance, nor can meaning be outsourced to "external forces" or "the way the world works." If the alignment begins with you, then what unfolds is not merely something that happens *to* you—it is something that happens *through* you.

This realization shifts perception from reactive to generative. You are not simply adjusting to reality; you are actively composing it. And that means design becomes possible—not just in the engineering sense, but in the daily act of shaping perception. You can deliberately frame a conversation to invite trust instead of conflict, or choose to interpret a challenge as an opening rather than an obstacle. Under UED, these are not motivational slogans; they are literal field adjustments that alter what the handshake makes real.

With this agency comes an unavoidable ethical weight. You cannot un-create the fact that every interaction carries your signature. Even non-action leaves an imprint, because withholding alignment still shapes the field. This awareness makes blame-shifting obsolete—you cannot hide behind "I was just reacting" when you now understand that reaction itself is a form of creation.

In this light, accountability becomes liberation. If you authored the ripple, you can revise it. If you shaped the field into one form, you can re-shape it into another. The key is conscious alignment—an intentional awareness of how your beliefs, emotions, and attention vectors are influencing the pattern.

And this is where the transition becomes inevitable: if the ripple starts within you, then the "light" that illuminates the universe is not a distant star, nor an abstract force—it is the field coherence born in the silent core of your own awareness. Which leads us directly into next chapter, where *The Light Within* ceases to be a poetic metaphor and becomes the literal origin of every connection you have ever known.

Chapter 20: The Light Within

Closing reflections on self, silence, and the power of aligned intent

Before we begin talking about the *Light Within*, let us pause. Not to think. Not to analyze. But to feel — the quiet hum behind your thoughts. You may have raced through pages of black holes, ripple collapse, cosmic stillness, ATT, and field coherence, but the story was never about distant stars or unreachable phenomena. It was always about this—the part of you that moves nothing, yet changes everything.

Mainstream science told you the universe began with a bang. Philosophy told you it began with a thought. UED tells you: *It begins when you align.*

20.1 – The Field Inside the Observer

We have been taught to think of ourselves as *in* the universe, moving through it like tiny passengers on a vast, indifferent ship. The galaxies are "out there," the stars are "up there," the planets "over there." The story of existence, as we inherited it, begins in the external world and only later trickles down into our awareness.

But that is not how reality works under UED.

Your awareness is not a bystander to the universe. It is not a telescope receiving data from a faraway source. It is the origin point of the handshake between the field and the patterns you call "experience." Without that handshake, there is no star to look at, no room to stand in, no self to situate.

Think of the "outside world" as a library with infinite books—but no pages are visible until you open one. The book may be sitting there in potential, but its story does not exist for you until you create the alignment that allows it to be read. In the same way, the objects, events, and sensations you believe are "out there" are only activated—brought into play—when your field meets them in a coherent state.

Not Just Brain, Not Just Mind

This is not the same as saying "It's all in your head." The head—the brain—is just a local processing hub. Your *mind* is a more expansive interface, but even that is not the whole story. UED places you in a **field**, one that is not limited by skin or skull. This field is continuously in contact with the structured vacuum energy (SVE), the very substrate in which all ripples—light, matter, motion—arise.

When you sense something, you are not "picking up" a finished signal sent across empty space. You are participating in the co-creation of a ripple alignment. It's as if the "signal" is only half-formed until it meets your field, and your field completes it.

The Still Core

Inside you—beneath thoughts, memories, emotions—is a zone that does not move. It does not change when your mood swings. It does not speed up when you're in a hurry. It does not shrink when you're afraid. It is a silent, perfectly coherent center.

Most people never notice it, not because it is hidden, but because it is quiet. The noise of reactive thinking drowns it out. Yet the still core is the real *you*. The rest—your opinions, your history, your plans—are patterns dancing on its surface.

From Passive to Generative

Here is the shift that changes everything: In mainstream thinking, perception is passive. Light travels from an object, bounces into your eyes, gets turned into nerve signals, and your brain "registers" it. In UED, perception is generative. Your field pre-aligns to certain ripple patterns, and that alignment is the event we call "seeing." The "light" you think you are receiving is the completion of a handshake you initiated.

This is why two people can look at the same scene and see utterly different things—not because one is "wrong," but because each has brought a different alignment pattern to the encounter.

Everyday Illustration

You've experienced this in simple ways:

- You notice a certain type of car everywhere the week after you buy one.
- A friend points out details in a painting you never saw, even though you've looked at it a dozen times.
- You meet someone new, and your first impression colors every future interaction—until one day a single event shifts that impression, and suddenly they "look different."

In all these cases, the external scene didn't change. Your field alignment did. The "world" you see is filtered through the ripples you're ready to match.

The Takeaway

The universe you think you live in is not "out there." It's *inside* the alignment space of your own field. You are not moving through a fixed cosmic map—you are continuously generating your own version of it through your alignment with the SVE.

And when you grasp this—not intellectually, but viscerally—you begin to understand why the *Light Within* is not just a metaphor. It is the only source you have ever truly seen.

20.2 – Silence as the True Laboratory

The moment you accept that the field you carry is the true stage of the universe, the next question is obvious: *How do I study it?*

The answer is not in bigger telescopes, faster colliders, or more sensitive detectors. It is in silence—because silence is the only condition in which the noise of misalignment subsides enough for the still field to reveal itself.

Science Without Silence

Modern science is almost allergic to stillness. We rush to measure, to stimulate, to accelerate. We shine lasers, fire particles, shake detectors, and call it "observation." But the very act of disturbance often drowns the subtle harmonics we claim to seek.

It's like trying to hear the faint hum of a tuning fork while standing next to a jet engine. The hum is still there—it's simply lost in the roar we've created.

The irony is that ancient traditions knew this. Mystics, sages, and contemplatives developed entire disciplines to reach states of profound internal quiet—not to escape reality, but to *enter* it more fully. Under UED, that's not superstition; it's an elegant technology of the self.

What Silence Really Means

Silence here is not the absence of sound, but the absence of field clutter—the random, reactive ripples generated by overthinking, emotional turbulence, and sensory overload.

You can be in a quiet room and still be deaf to your own still core because your mind is screaming through a thousand unresolved alignments. Likewise, you can be in a bustling market and still touch the still field inside if your alignments are settled.

Silence, in the UED sense, is coherence without interference. It is the calibration phase before any handshake with reality.

Silence as Experimental Method

In a lab, you isolate a variable to study it. In silence, you do the same with your awareness.

Here's the sequence:

1. Still the reactive loop. Let thoughts pass without chasing them.
2. Notice the background hum. This is not imagination—it's your field's baseline vibration.
3. Hold attention there. You're not "waiting for" anything; you're simply staying in phase with the still point.
4. Allow ripple awareness. Subtle shifts will register—not as sensory events, but as felt alignments.

When done consistently, this is more revealing than any optical instrument, because you're not peering *at* reality—you're seated in the place from which it arises.

Why Silence Works

If light is not a courier but a handshake, and if the handshake happens only in alignment, then any turbulence in your field will distort the exchange. Silence removes the turbulence.

It's no different from cleaning a mirror before using it to observe the night sky. The stars were always there—you just couldn't see them clearly because of the smudges.

The Risk of Never Pausing

Without moments of silence, you live entirely inside inherited alignments—patterns absorbed from culture, media, habits, and old emotional residues. You may think you're "seeing the world," but you're only seeing the inside of a locked loop.

In such a loop, even breakthroughs are incremental at best—tinkering with what's already inside the cage rather than stepping outside it.

From Observation to Participation

Silence doesn't just clear the channel—it changes the game. When the clutter falls away, you stop feeling like a receiver of reality and start realizing you are a participant in its unfolding. Your alignments become more deliberate. Your interpretations carry more precision. The "outer world" starts to bend—not magically, but mechanically—to the coherence you hold within.

The True Laboratory

Your body is the test chamber. Your field is the apparatus. Silence is the controlled environment. And you—the still core—are both the observer and the observed.

When you grasp this, science is no longer something "out there" in universities or research labs. Every moment of still awareness is an experiment in cosmic engineering.

20.3 – Intent as the Field's Lens

Silence clears the glass. Intent focuses it.

If your field is the still ocean of the universe, then intent is the lens that decides which ripples you'll magnify, align with, or initiate. And like any lens, intent can clarify an image—or distort it beyond recognition.

Why Intent Is Not Just "Desire"

In common language, intent is confused with wanting something badly enough. Under UED, intent is the pre-alignment state you hold before any handshake with reality occurs. It's not the craving after the fact—it's the coherent readiness before the ripple moves.

Desire without alignment is just agitation. Intent without alignment is wishful thinking. True intent is when your field is already shaped for the handshake you want, so when it comes, no energy is wasted in "translation."

The Lens Metaphor Made Real

A lens doesn't create light; it redirects and refines it. Likewise, your intent doesn't *generate* the raw field—it channels it into a specific, usable form.

If your intent is scattered, you produce chromatic aberrations in your life: misinterpretations, misfires, half-results. If your intent is precise, reality arrives in crisp coherence—what people call "luck" or "manifestation" is simply a phase-perfect handshake.

Silence + Intent = Directed Reality

Silence (20.2) strips away noise so your intent isn't fighting static. Once silence is established, the tiniest calibration in intent can shift enormous outcomes because the handshake has no distortion.

In practice:

1. Enter silence until your field stops reacting.
2. Form the intent as a felt state, not a verbal wish. (You don't say "I want calm"—you *hold* calm.)
3. Stay aligned until external conditions begin to echo it.
4. Release—because holding too tightly reintroduces turbulence.

Everyday Demonstrations

- **Conversations:** Entering a meeting with genuine openness vs. guarded defensiveness changes the entire tone without a single spoken word.
- **Creative work:** Artists who "see" the finished piece before starting aren't fantasizing—they're pre-aligning to it.
- **Problem-solving:** Approaching a challenge with curiosity instead of fear reconfigures what solutions you can even perceive.

In all cases, the shift happens *before* the first observable move.

The UED View of Influence

Mainstream thinking says influence is about force—pushing harder, talking louder, imposing your will. UED says influence is about field precedence—being the first to set the alignment so all following handshakes naturally match it.

This is why truly aligned people seem to "effortlessly" attract situations, resources, or allies. It's not metaphysical—it's mechanical. The field they hold is simply the easier path for other fields to follow.

The Shadow Side of Intent

A distorted lens can still focus energy—just in destructive ways. When intent is shaped by fear, resentment, or unacknowledged bias, you can still get results—but they will carry the turbulence of their source. This is how cycles of conflict perpetuate: each side's misaligned intent forces the other into reactive alignments, creating feedback loops of damage.

The responsibility of being the source (19.4) means you can't claim neutrality—your field is always shaping something, whether you like it or not.

From Lens to Light

Once you understand intent as the focusing lens of the still field, you stop seeing it as "positive thinking" or "goal setting." You start treating it like optical engineering:

- Adjusting curvature (clarity of purpose)
- Controlling aperture (breadth vs. precision)
- Removing distortion (emotional residue)

Because if the light within you is infinite, the only question is: How are you focusing it?

20.4 – Living as a Field Engineer

You've heard of software engineers, civil engineers, genetic engineers. A field engineer, in UED terms, works with something far more fundamental—the structured vacuum field that carries every ripple of reality.

And here's the twist: you've always been one. You just didn't know it.

What It Means to Engineer a Field

In the UED framework, "engineering" doesn't mean *building from scratch*—the field already exists, still and infinite. It means:

1. Recognizing the current state of your field.
2. Recalibrating its alignments.
3. Holding those alignments long enough for coherent outcomes to emerge.

This is a conscious shift from reactive living (being shaped by random ripple hits) to proactive living (shaping the ripple handshake you want before it arrives).

The Three Tools of a Field Engineer

1. **Silence** – Your calibration chamber. Without silence, you can't hear the field's baseline hum. Noise drowns out the subtle misalignments.
2. **Intent** – Your precision lens. Once the noise is gone, you can focus energy into specific, coherent forms.
3. **Observation** – Your feedback loop. You watch—not to judge success or failure, but to detect drift from your set alignment.

This trinity is not mystical; it's mechanical.

Daily Life as Field Work

Being a field engineer doesn't require a lab coat or quantum devices. It requires awareness at all scales:

- **Morning:** Before checking your phone, set your field with a single coherent intent. ("Today, I meet urgency with clarity.")
- **Meetings:** Notice when the group field starts drifting into turbulence. Adjust your tone, pace, or silence to reset alignment.
- **Creative work:** Treat each step as a handshake—don't move forward until you feel phase lock with the outcome.
- **Conflict:** Instead of "winning" the exchange, aim to restore the shared field to coherence. The solution often emerges automatically.

Field Drift and Realignment

Even the best-calibrated fields drift—emotions spike, unexpected inputs arrive, fatigue lowers coherence. The key is catching drift early:

- A sudden frustration is a phase shift—pause before acting.
- Feeling "out of sync" is an ATT fluctuation—reset through silence, not force.
- Over-attachment to an outcome creates lens distortion—loosen grip, let the field self-correct.

Scaling Up: From Self to Systems

When multiple individuals act as conscious field engineers, alignment scales:

- Teams find fewer misunderstandings and higher creative flow.
- Communities experience stability without heavy control

structures.
- Nations can shift from reactionary politics to anticipatory governance.

At the cosmic level, the same principle applies—large-scale coherence can make even the illusion of motion feel harmonious instead of chaotic.

Why This Changes Everything

Once you see yourself as a field engineer, you stop living in effect mode—the victim of causes "out there." Instead, you live in alignment mode, where:

- Perception precedes experience.
- Intent precedes manifestation.
- Stillness underlies every movement.

From this position, "The Light Within" is no longer poetic—it is operational truth. You are both the lamp and the lens, the still ocean and the ripple-maker.

Final Reflection

If you've made it this far, you've already begun engineering. Every pause you've taken to consider these ideas has been a micro-calibration. Every shift in how you see light, motion, or observation has already altered your handshake with the field.

And so the last question of the book is not, *"What will you do with this knowledge?"* It is: "What will you align with—now that you know you are the one doing the aligning?"

Closing Reflection – "The Light Within"

The journey through this book began with dismantling certainties—light, motion, time, matter—and may have left you with the unsettling sense that the universe is far less like the story you were taught, and far more like a silent stage where every scene is written in your own presence.

But now, at the end, all those dismantled concepts point to one truth so simple it almost hides in plain sight: The light you have been chasing was never "out there." It has always been inside.

Stillness Was Never Empty

We redefined light not as a courier racing across the void, but as a handshake—a local alignment of the structured vacuum field. We saw that motion was an illusion, time was a perception of ripple refresh, and space was a map of coherence relationships, not an expanse to be crossed. And once the noise of old narratives fell away, what was left? Not darkness. Not absence. But a field quietly ready to align—waiting for your intent to give it shape.

The Observer's Lamp

Throughout history, people have looked to the heavens for guidance—stars as fixed points, light as the great messenger. But under UED, the real fixed point is not the star—it's you. You are the lamp that lights the scene you perceive. You are the lens that sharpens or distorts it. And you are the storyteller who decides whether what you see is a threat, an opportunity, or a miracle.

This is not a metaphor. When you align your field, you literally determine what patterns can come into coherence with you. Your "light" is not an inner feeling—it is the operational principle of how reality meets you.

The Responsibility of Illumination

If the light is within, so is the shadow. When you distort alignment—through fear, prejudice, or restless distraction—the field returns that distortion to you. Not as punishment, but as the exact handshake you offered. Knowing this is both empowering and sobering. It means you can stop blaming the void for what you meet. It also means you can no longer claim ignorance for the ripples you launch.

This is the quiet responsibility of being the source: Every moment is a broadcast. Every alignment is an instruction to reality.

Silence as the Gateway

In a noisy mind, the light flickers, scattered by turbulence. In silence, the light stabilizes—its beam coherent, its field steady. This is why stillness has been revered by sages, scientists, and creators alike—not as an escape from life, but as the engineering bench where the next alignment is set.

The Light That Doesn't Fade

Stars burn out. Galaxies drift into invisibility. Technologies fail. But the light within—the capacity to align with reality at its most fundamental level—does not diminish. It is not powered by fuel, nor bound to time. It is the same light that existed before you knew your own name, and it will be the same long after you have forgotten it again.

Leaving the Reader with the Final Alignment

You began as an observer. You learned you were the source. Now you see you are also the light that makes observation possible.

From here, every ripple you meet will carry a question back to you: "Is this the alignment you choose?"

And when you answer—not with words, but with the quiet steadiness of your field—you will feel something subtle shift. Not in the stars. Not in the galaxies. Not even in the so-called outer universe.

It will happen where it has always happened—in the silent, still, luminous center of the field you call you.

Epilogue: The Field Is Yours Now

The journey has come full circle. We began with the light "out there," chasing photons across imagined distances, trusting stories of motion, expansion, and collapse. But with each chapter, the illusion peeled away—revealing not a restless, racing universe, but a still field quietly shaping everything you experience.

If this book has done its job, you now see that "light" was never just about stars, nor was it ever merely a physical quantity. It was the handshake between you and the field. And once you understand that the handshake begins within, the entire architecture of reality changes. The observer is no longer a passive recorder of the universe; the observer is its author.

Technology may still hum with old assumptions. Telescopes may still hunt for distant truths. Equations may still be written as if the cosmos were running somewhere. But knowing what you now know, you stand outside that story. You see the stillness beneath the motion, the coherence beneath the noise, the alignment beneath the spectacle.

And now comes the most difficult part—not believing, but living as if this is true. Every interaction becomes a ripple you set in motion. Every perception is a choice of alignment. Every moment is a chance to refine the coherence you bring to the field.

The universe will not move for you. It doesn't need to. It has always been here, waiting—still, whole, and ready for your intent.

The light you once thought you were chasing was never running away. It was always yours to begin with.

Appendices & Tools

Appendix A: UED vs Mainstream Optics

Comparison Table

This table outlines key conceptual differences between Mainstream Optics—rooted in photon-based, wave-particle duality assumptions—and UED Optics, which redefines light as a *field alignment handshake* rather than a traveling courier.

Theme	Mainstream Optics View	UED Optics View
Nature of Light	Dual entity—sometimes a wave, sometimes a particle—moving at finite speed through space.	Not an entity at all, but a *coherence handshake* between observer field and source field, decoded as "light."
Speed of Light	Constant universal velocity limit (*c*), independent of source or observer motion.	Not a velocity—*c* is the maximum refresh rate of field alignment (ATT limit), not a transit time.

Theme	Mainstream Optics View	UED Optics View
Photon Concept	Discrete packets of electromagnetic energy emitted, traveling through space, absorbed elsewhere.	A mathematical convenience; no particle crosses the void—only ripple-phase synchrony shifts.
Wave Behavior	Light behaves as an electromagnetic oscillation propagating through a medium-less vacuum.	Wave patterns are ripple refresh structures in SVE; no propagation—only local reconfiguration of alignment zones.
Transmission	Energy physically travels from emitter to receiver via photons or EM waves.	Energy is never "sent"—both points align within the same universal field structure.
Redshift	Caused by Doppler motion (velocity) or cosmic expansion stretching wavelengths.	A delay or drift in ripple alignment timing—no motion or space expansion required.

Theme	Mainstream Optics View	UED Optics View
Reflection/Refraction	Changes in photon direction/speed due to surface interaction or medium density.	Shifts in local field coherence and ATT thresholds—light paths are reassignments, not deflections.
Optical Limits	Diffraction, dispersion, and absorption are caused by wave interference, scattering, or medium resistance.	Limits emerge from alignment delay, misalignment noise, and ripple phase disruptions.
Causality	Light arrival marks when the emission event *actually happened* in the past.	Observation is a co-created event—alignment defines both the "when" and "where" simultaneously.
Experimental Aim	Measure and model light as a traveling energy form.	Map and control field alignment to engineer perception itself—lossless optics possible.

Appendix B. Core Ripple Concepts for Light

Understanding light under **Universal Energy Dynamics (UED)** requires abandoning the idea of *something traveling* and instead embracing the notion of **field alignment events**. Four key ripple-based concepts form the backbone of UED's optics framework:

1. Structured Vacuum Energy (SVE)

- **Definition:** The universal still-field that holds all potential alignments. Not empty space, but a structured, tension-bearing medium in absolute stillness.
- **Role in Light:** Light is not a separate "thing" moving through SVE. Instead, SVE reconfigures its local ripple phases so that two points—the observer and the source—share an aligned state.
- **Analogy:** A still pond where a fish and a bird appear to "see" each other instantly—not because anything traveled, but because the water's ripple pattern changed to connect them.

2. Alignment Threshold Time (ATT)

- **Definition:** The minimum time interval required for the SVE to update and stabilize a new ripple alignment between two points.

- **Why It Matters:** ATT is what mainstream physics mislabels as the "speed of light." Instead of a transit time for photons, it is the **refresh limit** for alignment changes in the field.
- **Consequences:**
    - Faster-than-light travel is meaningless—the refresh limit is a property of the universal field, not a barrier to be broken.
    - Variations in ATT across regions create illusions of acceleration, deceleration, or even cosmic expansion.

3. Ripple Delay

- **Definition:** The measurable lag between an event at the source and the observer's alignment to that event's ripple phase.
- **Distinction from ATT:** ATT is the *minimum possible* update time; Ripple Delay includes real-world misalignments, phase noise, and environmental interference.
- **Examples:**
    - **Astronomy:** Apparent "light-years" are not distances traveled but cumulative ripple delays from intermediate misalignments.
    - **Communications:** Signal lag is often due to phase recalibration limits, not transmission distance alone.

4. Alignment Zones

- **Definition:** Local regions in SVE where ripple phases

remain stable enough to support consistent decoding of an event.

- **Role in Perception:** Movement between alignment zones changes what an observer can decode—producing the illusion of motion, displacement, or even the birth/death of light sources.
- **Engineering Implications:**
    - Instead of trying to boost "signal strength," engineers could manipulate zone stability to maintain perfect alignment.
    - Imaging systems could bypass diffraction and dispersion limits by designing directly for zone coherence.

In short:

SVE is the still stage, ATT is the refresh clock, Ripple Delay is the lag in choreography, and Alignment Zones are the sections of the stage where the performance stays visible.

Appendix C. Glossary of Terms

A curated list of key concepts introduced or redefined under Universal Energy Dynamics (UED) for understanding light, perception, and cosmic stillness.

Chavan's Law

"No observation occurs without alignment; no alignment occurs without consequence."

A foundational UED principle stating that perception is not passive reception of pre-existing reality. Every observation is the result of a field alignment event in the Structured Vacuum Energy (SVE), and every alignment leaves a measurable consequence in ripple dynamics.

Observer Field

The localized ripple field generated by an observer's presence, intent, and decoding apparatus (biological or artificial).

- Acts as a *bridge* between universal field patterns and subjective perception.
- Determines what portion of the SVE's potential configurations can be accessed or "seen."

Ripple Response

The change in SVE's local configuration triggered by an event, resulting in a measurable shift in ripple phase alignment.

- In mainstream optics: mistaken for photons "arriving" at a detector.
- In UED optics: understood as a field handshake—the observer's field and the event's field coming into phase synchrony.

Structured Vacuum Energy (SVE)
A still, tension-bearing field permeating the universe.

- Not empty space; it contains the structured potential for all alignments.
- All motion, light, and causality are apparent—the SVE itself never moves.

Alignment Threshold Time (ATT)
The minimum interval required for the SVE to update and stabilize a new alignment between observer and source.

- Misinterpreted by mainstream physics as the "speed of light."
- Determines the refresh rate of reality as perceived by an observer.

Ripple Delay
The total lag between an originating event and the observer's field decoding that event's ripple phase.

- Includes ATT + interference effects + local misalignments.
- Explains astronomical "light travel time" without requiring photons to traverse space.

Alignment Zones

Stable regions in the SVE where ripple phases maintain coherence over time.

- Moving between zones can make objects appear to move, vanish, or distort without physical displacement.

Field Coherence

The degree to which ripple phases remain synchronized across an alignment zone.

- High coherence = sharp, stable perception.
- Low coherence = blurred, noisy, or inconsistent decoding.

Perceptual Handshake

The moment when the observer's field and the source's ripple pattern match phase well enough to decode information.

- This is the *real event* in seeing, hearing, or measuring—not a photon or particle exchange.

Phase Drift

Gradual misalignment between observer and source ripple phases, leading to apparent motion, redshift, or signal fading.

Ripple Horizon

The furthest limit from which an observer's field can maintain coherent alignment under existing ATT and ripple delay conditions.

- Analogous to the "observable universe" in mainstream cosmology, but determined by field coherence limits rather than distance.

Appendix D. Experimental Suggestions for Future Physics

Universal Energy Dynamics (UED) offers a radically different view of light—not as a courier crossing space, but as the result of field alignment events in the Structured Vacuum Energy (SVE). Testing this requires experiments that bypass conventional "photon travel" assumptions and focus instead on detecting, manipulating, and mapping ripple alignments.

These proposals are designed to be achievable with current technology, scalable for future refinement, and capable of producing results that directly challenge mainstream interpretations.

1. Field-Mapping Without Motion Assumptions

Goal: Detect alignment changes independent of any assumed object displacement.

- **Method:** Use an array of ultra-stable interferometers aimed at a fixed celestial target. Track phase shifts over time without calculating velocity or distance.
- **Expected UED Result:** Variations in recorded phase patterns will not correspond to any measurable physical movement of the source, but to observer-field realignments.
- **Challenge to Mainstream:** Mainstream optics will interpret these shifts as parallax or object motion—UED predicts measurable shifts even when no motion is possible.

2. ATT Measurement in the Lab

Goal: Directly measure the Alignment Threshold Time (ATT) for light-like events.

- **Method:**
    1. Create two coherent light sources with perfectly synchronized modulation.
    2. Introduce controlled phase offsets between source and detector fields.
    3. Measure the time required for phase realignment to stabilize detection.
- **Expected UED Result:** ATT will present as a minimum refresh interval, consistent regardless of the source's apparent velocity or distance.
- **Challenge to Mainstream:** Mainstream models have no parameter for a universal refresh rate—they expect signal timing to be purely distance-dependent.

3. Ripple Delay Interference Tests

Goal: Separate ripple delay effects from physical scattering or absorption.

- **Method:**
    1. Transmit a coherent signal through a medium of adjustable refractive index.
    2. Simultaneously record detection changes across multiple spatially-separated sensors.
    3. Compare timing differences to baseline vacuum tests.
- **Expected UED Result:** Variations will follow alignment delay curves, not standard refractive path-length predictions.

- **Application:** Could lead to lossless communication by designing for ripple alignment rather than photon transmission.

4. "Motion Stop" Event Capture

Goal: Detect moments where apparent motion halts due to field realignment.

- **Method:**
    1. Monitor a known high-proper-motion star using long-baseline interferometry.
    2. Look for brief intervals where proper motion appears to pause or reverse.
- **Expected UED Result:** Temporary stillness events will coincide with local alignment zone transitions—not with any change in the star's actual motion.
- **Challenge to Mainstream:** Conventional astronomy cannot explain sudden reversals without invoking extreme gravitational lensing or unknown forces.

5. Astronomical Fixed-Point Verification

Goal: Test the stillness hypothesis by anchoring observations to the Cosmic Microwave Background (CMB).

- **Method:**
    1. Map fixed CMB alignment points over extended time.
    2. Compare positional stability with reference stars or galaxies.
- **Expected UED Result:** CMB points remain fixed even when foreground objects appear to move—indicating motion is a decoding artifact, not a physical displacement.

6. Light Without Distance
Goal: Generate a detection event without a "travelling" photon.

- **Method:**
    1. Create synchronized ripple phase patterns in two separate, unlinked cavities.
    2. Induce identical field alignment conditions in both.
    3. Detect a change in one cavity triggered by alignment change in the other.
- **Expected UED Result:** Perception/detection occurs instantaneously within ATT limits—no measurable travel time.

7. Engineering Applications Testbed
Goal: Use UED principles to design optical systems that outperform mainstream efficiency limits.

- **Method:** Build a testbed that compares conventional fiber-optic transmission to a UED-inspired phase-alignment transmission system.
- **Metric:** Measure loss, bandwidth, and delay.
- **Expected UED Result:** Reduced signal loss and latency due to bypassing "photon transport" bottlenecks.

Closing Note: These experiments are not just validation attempts—they are invitation points for a new class of physics where understanding precedes engineering. The true breakthrough won't be a better telescope or faster laser, but a complete recalibration of how we think light, space, and observation work.

Acknowledgements

This work exists because reality refused to sit quietly in the form that textbooks offered it.

First, my gratitude to the skeptics—both the polite and the merciless. You questioned, doubted, dismissed, and sometimes even mocked these ideas. You were necessary. Every challenge forced a sharper articulation, a cleaner dismantling of the assumptions that have shaped mainstream optics for centuries. Without your resistance, this framework might still be an unspoken curiosity in the back of my mind.

To my students—you are the reason I kept peeling back the layers. Your questions never stopped at the edges of the syllabus, and your ability to see beyond the "official answer" reminded me that real learning is a rebellion against intellectual laziness. Many of the metaphors and explanations in this book began as classroom improvisations in response to your puzzled stares.

To my peers in science and engineering, even those still holding tightly to the photon-as-courier model—you have built the stage from which this conversation begins. Your work, whether in agreement or opposition, is part of the bridge between what we call "physics" and what physics might yet become.

To the silent field itself—the still, patient presence that reveals its nature only when approached without prejudice. This book is not an invention; it is a transcription of patterns you have been broadcasting since the first ripple disturbed the infinite ocean.

And finally, to those who will take this further—the ones who will run the experiments, publish the data, and either prove or refine this model. Whether you confirm every word or overturn half of it, your work will make the truth clearer. And that is the only victory that matters.

About the Author

Sandeep Chavan is an author, independent researcher, and educator whose work challenges some of the most deeply embedded assumptions in modern science. With a background in mechanical engineering and decades of experience teaching physics, chemistry, and mathematics, he has developed a reputation for dissecting complex ideas down to their core, then rebuilding them into frameworks that are both logically rigorous and intuitively clear.

Sandeep is the creator of **Universal Energy Dynamics (UED)**—a paradigm-shifting framework that redefines motion, light, time, and force as consequences of ripple alignment in a structured vacuum field (SVE). His books and research reports span topics from fundamental physics to human consciousness, bridging the gap between hard science, philosophy, and real-world application.

Never content with repeating accepted theories, he approaches knowledge like a patient but relentless investigator—testing, discarding, and refining until the explanation fits both the evidence and the deeper logic of reality. His classroom, whether under a roof or under a tree, has always been a place where "why" is more important than "what," and where curiosity is rewarded over conformity.

His work has attracted a global readership, from students seeking clarity to researchers and thinkers drawn to his unconventional but precise reasoning. He has authored over seventy books, including scientific treatises, philosophical explorations, and narrative works that bring his theories to life in unexpected ways.

When he is not writing or teaching, Sandeep continues to refine UED through field observations, thought experiments, and open debates with scientists, philosophers, and anyone brave enough to challenge him. He believes that truth is never owned—it is only borrowed from the field, tested, and passed on for the next mind to refine.

Final Note to the Reader

"You never saw light. You saw the universe looking back at you."

What you called "light" was never a stream of particles racing toward you. It was never a messenger crossing the void. It was your own field, your own alignment, meeting the field of what you believed was "out there" — and making the handshake complete.

Every sunrise you have ever witnessed, every starlit night that made you feel small, every flicker of a candle in the dark — they were not journeys of photons through unimaginable distances. They were moments when your awareness tuned to a ripple that had always been present, and in that tuning, reality bloomed.

This book was never about replacing one theory with another. It was about handing you the keys to your own observation, showing you that what you "see" is inseparable from who you are. The light you thought you were chasing was never outside of you — it was the alignment within you that made the world appear.

Carry this with you: the universe is not a stage on which you stand; it is a still ocean in which your every thought, intent, and observation creates the ripples you call reality. The next time you open your eyes and say, *"I see light,"* remember — the universe is looking back, through you, with you, as you.

Don't miss out!

Visit the website below and you can sign up to receive emails whenever Sandeep Chavan publishes a new book. There's no charge and no obligation.

https://books2read.com/r/B-A-EPGPC-UASTG

BOOKS 2 READ

Connecting independent readers to independent writers.

Did you love *Light: Neither a Wave, Nor a Photon*? Then you should read *Chavan's Declaration*[1] by Sandeep Chavan!

[2]

Chavan's Declaration *What if everything you've ever believed—about science, spirituality, and even your own self—is just a ripple on a deeper, structured field?*

Chavan's Declaration is not merely a book. It is a philosophical earthquake. A bold unification of ancient wisdom and radical science, it introduces a revolutionary framework: **Universal Energy Dynamics (UED)**—a ripple-based understanding of energy, consciousness, and reality itself.

This is not about rejecting science or worshipping mysticism. It's about decoding what both have been trying to express all along—using different waveforms. Beneath matter and mind lies

1. https://books2read.com/u/47pp2L

2. https://books2read.com/u/47pp2L

Structured Vacuum Energy (SVE)—a still field where all motion, form, and perception are ripple resolutions.

From gravity to God, from death to thought, from time to truth—this book dismantles outdated assumptions and replaces them with ripple logic:

The Chavan Ripple Field Equation (CRFE)—a proposed upgrade to Einstein's Field Equations**Mind, soul, and consciousness**—redefined as ripple behaviors, not metaphysical mysteries**Time**—not a flowing dimension, but a ripple consequence of causal tensions**Black holes**—not pulling matter, but resolving field imbalance**Religion, science, and scriptures**—recast as attempts to describe the same underlying alignment**AI vs. Human Awareness**—a ripple-level difference, not a computational one**And the final declaration:** Truth is not fixed—it's functional.

Whether you're a seeker, a skeptic, a teacher, or a physicist—this book doesn't ask you to agree. It invites you to **align**. To see not with belief, but with coherence.

Chavan's Declaration is the culminating volume of *The War of Thoughts* series—a philosophical rebellion against dogma, and an intellectual invitation to surf the ripple of truth.

The universe isn't broken. Only our explanations are.Dare to update them.

Read more at https://www.sandeepjchavan.com/.

Also by Sandeep Chavan

A Monk on World Tour
The Making of the Monk

Chavan's UED research
CMGED Research Report
Chavan's Cosmic Continuum (C^3)
Frequency
Ripples of Reality
Reality Realigned
The Great Realignment
I – The Cosmic Leverage
Chavan's Cosmic Theory of Relativity (CTR)
Chavan's Theory of Light
Feeling Like a Physicist
TIME – The New Definition
$E \neq mc^2$: The Collapse of Classical Assumptions
The Chavan Codex
Chavan's Atomic Model
Chavan's Theory of Everything

Echoes of War: The Russia-Ukraine Conflict and Its Global Ripples

Shattered Lives

The War Economy

Borders Redrawn

Arms and Allies

Collateral Damage

Justice in the Ashes

Somewhere in the Darkness of the Cosmos

The Blueprint of the Universe

The Cosmic Breath

From the Edge of the Universe

The Eternal Sage Chronicles

The Sage and the Sovereign

The Family Pillar

The People's Flame

The Warrior's Soul

The Bridging Mentor

The Guiding Light

The Exodus

The Last Cycle: 2075

The Fundamental Theory of Life & Existence

Energy in Action: The Soul Demystified

The Active Mind

Energy in Connection

Death: The Disconnection

Subconsciousness: The Chosen Disconnection

Desires: The Engine of Life

The Restoration of Human Knowledge

Restoring Bhagavad Gita

The War of Thoughts

Cracks in the Known

The Human Game

The Divine Dilemma

The Original Intent

The Answers

Ultra Modern Life Philosophy

The Duality of Laughter and Tears

Take It Easy

Compare to Accept

Standalone

It's Not AI, It's AHI - Amplified Human Intelligence
The Decision Paradox
The Four Sapiens
Malicious Script of Indian Polity
The IIT Legacy & Global Impact
The IIT Legacy & Global Impact
You are just a Version of Your Original
Win the Game You Didn't Choose
Always Keep Your Bags Packed
Reveal to Shield - Learn the Game of Facades
Engage Beyond Elections
Culture, Identity & Change: The Evolution of Indian Society
The Triangular Dynamics
Never Trust Without Doubt
If You Can't Change the Country, Leave the Country
Live Fearless with Fear
India: The Investment Magnet
Realignment
Powering a Billion Dreams
Universal Skill
Extension: From Brush to Algorithm
Human Side of Einstein
The Future Knows No Lines
Let's Bring the Generations to the Table
Unified Energy Dynamics
A Beat That Never Dies
Strategic Horizons
Right From Scratch
The Forgotten Innovators
Physics Fixing
The Savior Illusion
The Selfish Side
The Age of the Weakest Sapiens

Neural Wars
The Absorption: There Is No Death
Strange Therapist
Bubbles: Become a Drop
Beyond the Blue
Half-Scholar
The Law of Truth
Failure
Philosophy 2.0: A Structural Manifesto
Chaos: A Gentle Guide to the Unresolvable
Self-Help: New Syllabus
Rippling
Game: How to Play the Share Market
Out of Syllabus
Saffronized: The Code Behind the Crown
The Anger
Moksha: The Liberation
Chavan vs Einstein: The Universe Was Never Curved
Rocket Science: The Ripple Gambling
The Cosmos: The Still Universe
The Taste of Silence
Chavan's Post-Materialist Human Model
10-Second Rule
Chavan vs Hawking
It was Never a Wave Signal
Quantum Courtroom: The Trial of a Century
Chavan–Heisenberg Ripple Uncertainty Principle
The Great Disruption
Dark ≠ Hidden
Breath
Calculus of Life
The Earth Doesn't Move
Black Hole Don't Exist

Chavan vs Kant
Naked Truth: Everything and Else
The Moon: Co-Author of Life's Script
World War-3
Aham Brahmasmi
Chavan's Declaration
Light: Neither a Wave, Nor a Photon
Chavan vs. Plato

Watch for more at https://www.sandeepjchavan.com/.

www.ingramcontent.com/pod-product-compliance
Lightning Source LLC
LaVergne TN
LVHW090310210926
848480LV00001B/191
* 9 7 9 8 2 3 1 0 9 0 2 1 1 *